How to Afford
Your Own Log Home

"The secret lies in being your own contractor—and the author should know because he is a contractor, specializing in log homes."

—**Popular Science** magazine

"Not just for Lincoln nuts, this book explains that contemporary log homes combine energy efficiency and practicality with nostalgia for a past way of life."

—**The Philadelphia Inquirer**

"Different phases of building are covered with an eye toward minimizing the red tape a beginner might encounter in the building industry. Sample contracts are included for easy reference, while sections listing the names, addresses, and phone numbers of precut log home manufacturers complete the invaluable information to potential log home purchasers."

—**The Midwest Book Review**

"I wish we'd had this book before we built our log home."

—Doris Muir, Publisher
**Log Home Guide for
Builders and Buyers**

Help Us Keep This Guide Up to Date

Every effort has been made by the author and editors to make this guide as accurate and useful as possible. However, many things can change after a guide is published—establishments close, phone numbers change, facilities come under new management, etc.

We would love to hear from you concerning your experiences with this guide and how you feel it could be made better and be kept up to date. While we may not be able to respond to all comments and suggestions, we'll take them to heart and we'll also make certain to share them with the author. Please send your comments and suggestions to the following address:

The Globe Pequot Press
Reader Response/Editorial Department
P.O. Box 480
Guilford, CT 06437

Or you may e-mail us at:

editorial@globe-pequot.com

Thanks for your input.

How to Afford Your Own Log Home

Save 25% without Lifting a Log

Fourth Edition

by Carl Heldmann

The Globe Pequot Press

Guilford, Connecticut

Cover Photo Credits
Front: Courtesy of Maple Island Log Homes.
Back: From top to bottom, Glen Graves/courtesy of Maple Island Log Homes; courtesy of Gastineau Log Homes, Inc.; courtesy of Real Log Homes; courtesy of Real Log Homes; and Glen Graves/courtesy of Maple Island Log Homes.

Cover Design by Saralyn D'Amato-Twomey.

Library of Congress Cataloging-in-Publication Data

Heldmann, Carl.
    How to afford your own log home: save 25% without lifting a log / by Carl Heldmann.—4th ed.
        p.   cm.
    Includes bibliographical references and index.
    ISBN 0-7627-0113-7
    1. Log cabins—Finance. 2. Log cabins—Design and construction.
    I. Title.
    HD7289.A3H44 1997
    643'.12—DC21   97-20131
                                                          CIP

Manufactured in the United States of America
Fourth Edition/Third Printing

Contents

About the Author

Carl Heldmann is a veteran licensed general contractor and was formerly involved in log home construction and sales. He is now a mortgage consultant in financing log home construction.

Heldmann is the founder of two schools of building, in North Carolina and in Michigan. He is the author of several books on construction.

Preface

The incredible growth in the popularity of log homes is quite understandable. Log homes are beautiful, reasonably priced, energy efficient, easy to build and maintain, very durable, a symbol of a person's spirit, and a spirit of, but a far cry from, log cabins of years gone by. The warmth of a log home is felt from both the outside and the inside. The richness of wood, our greatest renewable resource, is pleasing to the eye as well as to the pocketbook.

Study after study has proved that a log home is easier than most homes to heat in the winter and cool in the summer. Wood is a natural insulator! Studies show that it would take a concrete wall that is 5 feet thick to meet the insulating quality of only 4 inches of wood. Wood insulates 6 times better than brick, and an amazing 1,700 times better than aluminum. Why is wood such a good insulator? Wood is made up of millions of tiny air cells. These air cells act like tiny vacuum bottles slowing the transfer of heat in either direction, keeping heat in during winter and out during summer. Wood is also incredibly strong. Pound for pound, wood is stronger than steel. Log homes can, and do, last for centuries.

Some prospective log home owners would rather build with stud construction, either for purposes of using conventional insulation materials or for interior aesthetics. Some log home manufacturers, therefore, have made available log siding that makes a conventionally framed house look like a log home. Some manufacturers even provide the log corners to enhance this look. Either way, full log or half log, you can't go wrong with a well-designed and well-built log home. The manufacturers of today's log homes make it easier than ever to obtain and afford one. I hope you find the log home of your dreams, and I hope this book helps you afford it.

Beaver-tooth log ends give a rustic look to this home on Flathead Lake in Montana. Large wraparound decks with generous helpings of colorful flowers highlight this beautiful handcrafted house by Montana Log Homes.

Introduction

The subtitle of this book clearly states its purpose: to save 25 percent of the cost of your log home without your lifting a log. It is the purpose of this book to show you that you can avoid a down payment, lower the amount of mortgage needed, beat high interest rates, or get a larger log home for your money while saving 25 percent, plus or minus, and all without doing any physical work. You can! You can do this even if you have a full-time job. If you elect to do some of the work, your savings could be even greater.

This book will show you how to acquire a log home by being your own general contractor. A general contractor is the person who hires the people who will build your log home. You will see that being your own general contractor is not difficult and is very rewarding, both monetarily and personally. No license is necessary for building your own log home, nor is any professional knowledge of the construction trade. A log home kit, or in some cases an erected log shell, makes the job of being your own general contractor even easier. In most cases, you will have guidance from the log home manufacturer, another plus for you. You will also learn how to work with a professional general contractor, if circumstances require it, and still save thousands.

In this book you will also learn some of the attributes of log homes. To learn more about any particular log home, you will need to contact some of the many log home manufacturers listed with addresses in the back of this book. A glance at the table of contents will show you that all you need to know is covered in this book. The National Association of Home Builders estimates that hundreds of thousands of people just like you will build their own homes this year alone. The North American Log Builders Association estimates that tens of thousands of people will build their own log homes this year. You can succeed! Good luck!

For more information please visit my web site at www.byoh.com.

Note: Glossary terms are indicated by italic letters.

PHASE I
THE PLANNING STAGE

The Walton is a popular model offered by Amerlink. Its classic exterior lines house a spacious and functional floor plan. Available in white pine or western red cedar, the Walton provides 2,208 square feet of living area.

CHAPTER 1

Being Your Own General Contractor

Most people, upon hearing the term general contractor, conjure up a picture of a big burly guy hammering nails or laying brick. This may have been the case in years gone by, but it is rarely true today. Today's general contractor is more likely to be a manager of the time, people, and money that go into the construction of a structure than the one who actually does the work. He employs the professional people, called subcontractors, who actually construct the house. The term "builder" is very misleading, for a general contractor or builder usually builds nothing himself. He manages the people who do. For this management function, he is paid quite well. You will be, too, in the form of savings on the cost of your log home. Your savings can be used as your down payment to lower the amount of mortgage you will need, thereby effectively beating high interest rates if rates are up, or to get a larger log home for your money. Of course, you could elect to do some of the physical work yourself, thereby being a subcontractor as well, and save additional money.

There are other reasons for being your own general contractor. You and you alone will guide the construction project toward your final dream. You will be more assured of getting exactly what you want and at a price you can afford. You will control quality and cost.

How Much Will You Save?

How much you will save by being your own general contractor will vary with each individual. There are many factors that govern your savings: the size of your log home, the price you pay for your land, the cost of labor in your area, and the cost of the other materials that will go into your log home. A savings of 25 percent is not impossible. Keep in mind that whatever you save, you will NOT have to pay taxes on. This makes your savings even greater. NOTE: It's an even greater savings when you realize that you won't have to pay interest or repay principal on the amount you save.

A typical example of savings is shown on the next page. As you will see, all financing figures are based on the appraised value, also called market value, of an existing or proposed home. The following is based on a 1,500-square-foot (heated area) log home in Grand Rapids, Michigan.

(Courtesy of Mountaineer Log Homes)

A home by Mountaineer Log Homes under construction.

Costs are for 1997 and are approximate. They may be greater in some areas but the percentages should stay the same.

EXAMPLE:

1,500 sq. Ft. heated area @ $106/sq. ft.

Retail, including real estate commissions = $160,000

Actual total cost of construction = $90,000

Land cost = $30,000

Total cost of log home = $120,000

Your savings as general contractor = $40,000

This savings, which includes the real estate commission that you would pay, is equal to approximately 25 percent of the appraised (market) value of this log home. As you will see in chapters 5 and 6, you will be able to borrow the full $120,000 for construction financing and the permanent loan, or mortgage. This means that you will have none of your own money invested in your dream log home . . . NO DOWN PAYMENT! Or, if you have money for a down payment, you could use that money to lower the amount you would need to borrow.

What Do You Need to Know?

The only major tools necessary for being your own general contractor are a telephone, a calculator, and a checkbook. You need only to know how to organize your spare time and treat people in a fair manner, and you are in business. If that sounds simple, it is. You won't have to have technical knowledge of any of the fields of subcontracting, since your subcontractors are hired for their expertise in their fields. You don't have to know anything about plumbing to call a plumber to fix a leak or replace a defective plumbing fixture. The same holds true when you are building your own home. Your role as general contractor is to manage time, people, and money. You already do this every day by managing the household budget, comparison shopping in the supermarket, balancing family time with work time, motivating people at work or at home, hiring repair people for various jobs, and so on.

You already have the knowledge to be your own general contractor. Keep this in mind throughout the process of building your home, especially when talking to lenders, and your confidence in yourself will show and be a help to you.

How Much of Your Time Is Required?

Consider the job of being a general contractor not as one large task but as a series of little jobs, for that is what it is. Each little job in itself does not require much of your time. All of them can be carried out in your spare time. In the planning, financing, and estimating phases, you can proceed at your own pace, at your leisure if you wish. The building phase can also be handled so as not to interfere with your job or family life.

Planning can be done in the evenings or on weekends. The same holds true for estimating. Carrying out the first three phases of being your own general contractor is called doing your homework. By working with a log home manufacturer, you'll have a lot of your homework done for you. As we go through each phase, this will become evident. Other people can help you with your homework, as you will see. A quick example is using a real estate broker

(Courtesy of Wisconsin Log Homes)

New Cottage Log™ hand-peeled siding produced by Wisconsin Log Homes in Green Bay can be used for interior or exterior application. The 2-inch-thick siding was spaced for chinking in this room. Cottage Log™ comes in 6- or 8-inch heights of northern white cedar or white pine and is shiplapped for easier application.

to check out all the things that need to be checked out before buying land. We will cover those in chapter 3. There are other helpers along the way. Some will help you free of charge, while others will charge a nominal fee. Any costs incurred are costs that all general contractors incur and are merely part of the cost of any home.

Do You Need a License?

You do NOT need a license to be your own general contractor for the purpose of building your own home. PERIOD! If you were to be a general contractor and construct a home for someone else, in most areas you would need a contractor's license and/or a business license.

Can You Get a Loan?

In chapter 5 on financing, you will see how to overcome the problems you might have in obtaining construction financing while acting as your own general contractor. This is mentioned now only because it is necessary to read this book thoroughly and more than once before you talk to prospective lenders.

Can You Do Your Own Labor?

If you are planning to do some of the physical work yourself or have friends do it, be sure to find out if any specific subcontractor license and/or permits are required. Most areas require plumbing, electrical, and mechanical (heating and air-conditioning) work to be installed by licensed subcontractors.

Do You Need Permits?

In most areas, you will need certain permits, but they are easy to obtain and are not costly. The purpose of permits is to enable a local government to set up and maintain an inspection department. This inspection department is then responsible for the compliance with certain *building codes* by all subcontractors. You will see that your local building inspection department is one of your helpers. If you don't have such a department in your locale, read chapter 8 to learn how you can hire your own inspectors.

If You Can't Be Your Own General Contractor

If for some reason, even after carefully reading this book, you feel that you will not be able to be your own general contractor, then there is an alternative that will still save you thousands of dollars. It is called a *manager's contract*. An example is shown at the end of chapter 1. It is only an example. You would need to have a real estate attorney draw up a contract that would protect you and be binding in your state. A manager's contract allows you to hire a professional general contractor to perform any part of the process of being a general contractor. The more you want him to do and the more responsibility you give him, the more it will cost you. If you want him only to do the building stage, with the exception of buying from suppliers, he probably would charge you about half what he would charge if he were to do all four

(Courtesy of Mountaineer Log Homes)

A newly completed log home by Mountaineer Log Homes.

phases. He would merely be "managing" your subcontractors for you. Of course, you could have him do more, but with an increase in his fee. You and a real estate attorney can make up your own manager's contract to suit your needs and in agreement with a professional general contractor.

There are two other forms of contracts at the end of this chapter, which show you two other ways to employ a professional general contractor. The "fixed fee" contract is usually the least expensive of the two but certainly not as inexpensive as a manager's contract, because the professional general contractor's responsibilities are increased with the fixed fee contract.

Manager's Construction Contract

1. General

This contract dated _____ is between

_____ (OWNER) and

_____ (MANAGER)

and provides for supervision of construction by MANAGER of a Log Home to be built on OWNER'S Property at _____ , and described as _____ .

The project is described on plans dated _____ and specifications dated _____ , which documents are a part hereof.

2. Schedule

The project is to start as near as possible to _____ , with anticipated completion _____ months from starting date.

3. Contract Fee and Payment

3A. OWNER agrees to pay MANAGER a minimum fee of _____

($) for the work performed under this contract, said fee to be paid in installments as the work progresses as follows:

a. Foundation complete	$	_____
b. Kit constructed	$	_____
c. Dried-in	$	_____
d. Ready for drywall	$	_____
e. Trimmed out	$	_____
f. Final	$	_____

3B. Payments billed by MANAGER are due in full within ten (10) days of bill mailing date.

3C. Final payment to MANAGER is due in full upon completion of residence; however, MANAGER may bill upon "substantial completion" (see paragraph 11 for the definition of terms) the amount of the final payment less 10 percent of the value of work yet outstanding. In such a case, the amount of the fee withheld will be billed upon completion.

4. General Intent of Contract

It is intended that the OWNER be in effect his own "General Contractor" and that the MANAGER provide the OWNER with expert guidance, advice, and supervision and coordination of trades and material delivery. It is agreed that MANAGER acts in a professional capacity and simply as agent for OWNER and that as such he shall not assume or incur any pecuniary responsibility to contractor, subcontractors, laborers, or material suppliers. OWNER will contract directly with subcontractors, obtain from them their certificates of insurance and lease of liens. Similarly, OWNER will open his own accounts with material suppliers and be billed and pay directly for materials supplied. OWNER shall insure that insurance is provided to protect all parties of interest. OWNER shall pay all expenses incurred in completing the project, except MANAGER'S overhead as specifically exempted in paragraph 9. In fulfilling his responsibilities to OWNER, MANAGER shall perform at all times in a manner intended to be beneficial to the interests of the OWNER.

5. Responsibilities of Manager

General

MANAGER shall have full responsibility for coordination of trades, ordering materials and scheduling work, correction of errors and conflicts, if any, in the work, materials, or plans, compliance with applicable codes, judgment as to the adequacy of trades' work to meet standards specified, together with any other function that might reasonably be expected in order to provide OWNER with a single source of responsibility for supervision and coordination of work.

Specific

1) Submit to OWNER in a timely manner a list of subcontractors and suppliers MANAGER believes competent to perform the work at competitive prices. OWNER may use such recommendations or not at his option.

2) Submit to OWNER a list of items requiring OWNER'S selection, with schedule dates for selection and recommended sources indicated.

3) Obtain in OWNER'S name(s) all permits required by governmental authorities.

4) Arrange for all required surveys and site engineering work.

5) Arrange for all installation of temporary services.

6) Arrange for and supervise clearing, disposal of stumps and brush, and all excavating and grading work.

7) Develop material lists and order all materials in a timely manner, from sources designated by OWNER.

8) Schedule, coordinate, and supervise the work of all subcontractors designated by OWNER.

9) Review, when requested by OWNER, questionable bills and recommend payment action to OWNER.

10) Arrange for common labor for hand digging, grading, and cleanup during construction and for disposal of construction waste.

11) Supervise the project through completion, as defined in paragraph 11.

6. Responsibilities of Owner

OWNER agrees to:

1) Arrange all financing needed for project, so that sufficient funds exist to pay all bills within ten (10) days of their presentation.

2) Select subcontractors and suppliers in a timely manner so as not to delay work. Establish

charge accounts and execute contracts with same, as appropriate, and inform MANAGER of accounts opened and of MANAGER'S authority in using said accounts.

3) Select items requiring OWNER selection and inform MANAGER of selections and sources on or before date shown on selection list.

4) Inform MANAGER promptly of any changes desired or other matters affecting schedule so that adjustments can be incorporated in the schedule.

5) Appoint an agent to pay for work and make decisions in OWNER'S behalf in cases where OWNER is unavailable to do so.

6) Assume complete responsibility for any theft and vandalism of OWNER'S property occurring on the job. Authorize replacement/repairs required in a timely manner.

7) Provide a surety bond for his lender if required.

8) Obtain release of liens documentation as required by OWNER'S lender.

9) Provide insurance coverage as listed in paragraph 12.

10) Pay promptly for all work done, materials used, and other services and fees generated in the execution of the project, except as specifically exempted in paragraph 9.

7. Exclusions

The following items shown on the drawings and/or specifications are NOT included in this contract, insofar as MANAGER supervision responsibilities are concerned:

(List below) _____

_____

_____

_____

_____

_____

_____

_____

_____

_____

_____

_____

8. Extras/Changes

MANAGER'S fee is based on supervising the project as defined in the drawings and specifications. Should additional supervisory work be required because of EXTRAS or CHANGES occasioned by OWNER, unforeseen site conditions, or governmental authorities, MANAGER will be paid an additional fee of 15 percent of cost of such work. Since the basic contract fee is a minimum fee, no downward adjustment will be made if the scope of work is reduced, unless contract is canceled in accordance with paragraphs 13 or 14.

9. Manager's Facilities

MANAGER will furnish his own transportation and office facilities for MANAGER'S use in supervising the project at no expense to OWNER. MANAGER shall provide general liability and workmen's compensation insurance coverage for MANAGER'S direct employees only, at no cost to OWNER.

10. Use of MANAGER'S Accounts

MANAGER may have certain "trade" accounts not available to OWNER which OWNER may find it to his advantage to utilize. If MANAGER is billed and pays such accounts from MANAGER'S resources, OWNER will reimburse MANAGER within ten (10) days of receipt of MANAGER'S bill at cost plus 8 percent of such materials/services.

11. Project Completion

1. The project shall be deemed complete when all the terms of this contract have been fulfilled and a Residential Use Permit has been issued.

2. The project shall be deemed "substantially complete" when a Residential Use Permit has been issued and less than five hundred dollars ($500) of work remains to be done.

12. Insurance

OWNER shall insure that workmen's compensation and general liability insurance are provided to protect all parties of interest and shall hold MANAGER harmless from all claims by subcontractors, suppliers and their personnel and for personnel arranged for by MANAGER in OWNER'S behalf, if any.

OWNER shall maintain fire and extended coverage insurance sufficient to provide 100 percent coverage of project value at all stages of construction, and MANAGER shall be named in the policy to insure his interest in the project.

Should OWNER and MANAGER determine that certain subcontractors, laborers, or suppliers are not adequately covered by general liability or workmen's compensation to protect OWNER'S and/or MANAGER'S interests, MANAGER may, as agent of OWNER, cover said personnel on MANAGER'S policies, and OWNER shall reimburse MANAGER for the premium at cost plus 10 percent.

13. MANAGER'S Right to Terminate Contract

Should the work be stopped by any public authority for a period of thirty (30) days or more through no fault of the MANAGER, or should work be stopped through act or neglect of OWNER for ten (10) days or more, or should OWNER fail to pay MANAGER any payment due within ten (10) days written notice to OWNER, MANAGER may stop work and/or terminate this contract and recover from OWNER payment for all work completed as a proration of the total contract sum, plus 25 percent of the fee remaining to be paid if the contract were completed as liquidated damages.

14. OWNER'S Right to Terminate Contract

Should the work be stopped or wrongly prosecuted through act or neglect of MANAGER for ten (10) days or more, OWNER may so notify MANAGER in writing. If work is not properly resumed within ten (10) days of such notice, OWNER may terminate this contract. Upon termination, entire balance then due MANAGER for that percentage of work then completed, as a proration of the total contract sum, shall be due and payable and all further liabilities of MANAGER under this contract shall cease. Balance due MANAGER shall take into account any additional cost to OWNER to complete the house occasioned by MANAGER.

15. MANAGER/OWNER'S Liability for Collection Expenses

Should MANAGER or OWNER respectively be required to collect funds rightfully due him through legal proceedings, MANAGER or OWNER respectively agrees to pay all costs and reasonable attorney's fees.

16. Warranties and Service

MANAGER warrants that he will supervise the construction in accordance with the terms of this contract. No other warranty by MANAGER is implied or exists.

Subcontractors normally warrant their work for one year, and some manufacturers supply yearly warranties on certain of their equipment; such warranties shall run to the OWNER, and the enforcement of these warranties is in all cases the responsibility of the OWNER and not the MANAGER.

(MANAGER) _____ (SEAL) DATE _____

(OWNER) _____ (SEAL) DATE _____

(OWNER) _____ (SEAL) DATE _____

Fixed Price Contract

CONTRACTOR: _____

OWNER: _____ DATE: _____

OWNER is or shall become fee simple owner of a tract or parcel of land known or described as: _____ . CONTRACTOR hereby agrees to construct a Log Home and the specifications herein attached.

OWNER shall pay CONTRACTOR for the construction of said house $ _____ .

Prior to commencement hereunder, owner shall secure financing for the construction of said house in the amount of $ _____ , which loan shall be disbursed from time to time as construction progresses, subject to a holdback of no more than 10 percent. OWNER hereby authorizes CONTRACTOR to submit a request for draws in the name of the OWNER from the savings and loan, or similar institution, up to the percentage of completion of construction and to accept said draws in partial payment hereof.

CONTRACTOR shall commence construction as soon as feasible after closing and shall pursue work to a scheduled completion on or before seven (7) months from commencement, except if such completion shall be delayed by unusually unfavorable weather, strikes, natural disasters, unavailability of labor or materials, or changes in the plans and specifications.

CONTRACTOR shall build the residence in substantial compliance with the plans and specifications and in a good workmanlike manner and shall meet all building code requirements. CONTRACTOR shall not be responsible for failure of materials or equipment not CONTRACTOR'S fault. Except as herein set out, CONTRACTOR shall make no representations or warranties with respect to the work to be done hereunder.

OWNER shall not occupy the residence and CONTRACTOR shall hold the keys until all work has been completed and all moneys due CONTRACTOR hereunder have been paid.

OWNER shall not make any changes to the plans and specifications until such changes shall be evidenced in writing, the costs, if any, of such changes shall be set out, and any additional costs thereof shall be paid in advance of the work being accomplished.

CONTRACTOR shall not be obligated to continue work hereunder in the event OWNER shall breach any term or condition hereof, or if for any reason construction draws shall cease to be advanced upon proper request thereof.

Any additional or special stipulations attached hereto and signed by the parties shall be and are made a part hereof.

CONTRACTOR: _____ (SEAL)

OWNER: _____ (SEAL)

OWNER: _____ (SEAL)

Fixed Fee Contract

CONTRACTOR: _____

OWNER: _____ DATE: _____

OWNER is or shall become fee simple owner of a tract or parcel of land known or described as: _____ . CONTRACTOR hereby agrees to construct a Log Home on the above described lot according to the plans and specifications identified as: Exhibit A—plans and specifications drawn _____ by _____ . OWNER shall pay CONTRACTOR for the construction of said house cost of construction and a fee of _____ . Cost is estimated in Exhibit B. Each item in Exhibit B is an estimate and is not to be construed as an exact cost.

OWNER shall secure/has secured financing for the construction of said house in the amount of cost plus fee, which shall be disbursed by a savings and loan or bank from time to time as construction progresses, subject to a holdback of no more than 10 percent. OWNER hereby authorizes CONTRACTOR to submit a request for draws in the name of OWNER under such loan up to the percentage completion of construction and to accept said draws in partial payment hereof. In addition, it is understood that the CONTRACTOR'S fee shall be paid in installments by the savings and loan or bank at the time of and as a part of each construction draw as a percentage of completion, so that the entire fee shall be paid at or before the final construction draw.

CONTRACTOR shall commence construction as soon as feasible after closing of the construction loan and shall pursue work to a scheduled completion on or before seven (7) months from commencement, except if such completion shall be delayed by unusually unfavorable weather, strikes, natural disasters, unavailability of labor or materials, or changes in the plans or specifications.

CONTRACTOR shall build the residence in substantial compliance with the plans and specifications and in a good and workmanlike manner and shall meet all building codes. CONTRACTOR shall not be responsible for failure of materials or equipment not CONTRACTOR'S fault. Except as herein set out, CONTRACTOR shall make no representations or warranties with respect to the work to be done hereunder. OWNER shall not occupy the residence and CONTRACTOR shall hold the keys until all work has been completed and all moneys due CONTRACTOR hereunder have been paid.

OWNER shall not make changes to the plans or specifications until such changes shall be evidenced in writing, the costs, if any, of such changes shall be set out, and the construction lender and CONTRACTOR shall have approved such changes. Any additional costs thereof shall be paid in advance, or payment guaranteed in advance of the work being accomplished.

CONTRACTOR shall not be obligated to continue work hereunder in the event OWNER shall breach any term or condition hereof, or if for any reason the construction lender shall cease making advances under the construction loan upon proper request thereof. Any additional or special stipulations attached hereto and signed by the parties shall be and are made a part hereof.

OWNER: _____ (SEAL)

OWNER: _____ (SEAL)

CONTRACTOR: _____ (SEAL)

CHAPTER 2

Selecting Your Log Home Kit

Selecting a log home is like selecting a new car. You will want to shop for what fills your needs. Cost, size, style, quality, ability to deliver on time, warranties, and references are some of the guidelines you should use in making your decision.

COST: Log home companies are very competitive in their pricing. In comparing costs, keep in mind the following: different kinds of wood, size of the logs (thickness), manner of construction, charges for plans, cost of freight to your job site, and what is included in the kit or package. This last item is, in my estimation, the most important. It is very difficult to compare "apples with apples" unless you know exactly what you are getting for your money. Will you be getting just the basic shell, or will you be getting everything necessary for *drying in* the house? Does the basic kit or shell include the roof system or just the rafters? Are the *subfloor* and its *framing members* included in the price? Can you buy the materials necessary for drying in the house locally for less? Will the log home company give you a list of these materials free of charge so that you can compare? In order to be fair to both you and the log home company, it is important that you ask these questions. You will find that company representatives are used to such questions and will be most helpful.

SIZE: We will discuss size fully in chapter 4, but size is singularly the most important factor in determining price. You will need to consider it carefully. Most companies can offer any size log home.

STYLE: This is a very personal decision. There is a tremendous variety of styles available, both within each individual company and among the various companies. You should have no trouble finding a style to suit you.

QUALITY: Today's log homes are a far cry from those of yesterday. You will find that the quality offered by most log home manufacturers is excellent. You should, if at all possible, visit a model home to inspect for quality.

DELIVERY: Most companies give guaranteed delivery dates. Be sure that they do. When your

Full Log Construction
(Courtesy of Greatwood Log Homes)

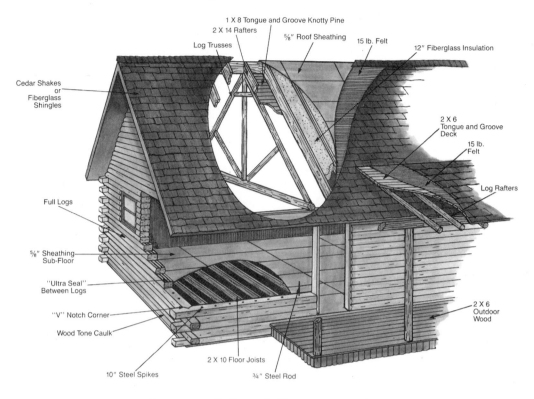

1 X 8 Tongue and Groove Knotty Pine
2 X 14 Rafters
⅝" Roof Sheathing
15 lb. Felt
12" Fiberglass Insulation
Log Trusses
Cedar Shakes or Fiberglass Shingles
2 X 6 Tongue and Groove Deck
15 lb. Felt
Log Rafters
Full Logs
⅝" Sheathing Sub-Floor
"Ultra Seal" Between Logs
"V" Notch Corner
Wood Tone Caulk
2 X 6 Outdoor Wood
10" Steel Spikes
2 X 10 Floor Joists
¾" Steel Rod

Log and Stud Construction
(Courtesy of Greatwood Log Homes)

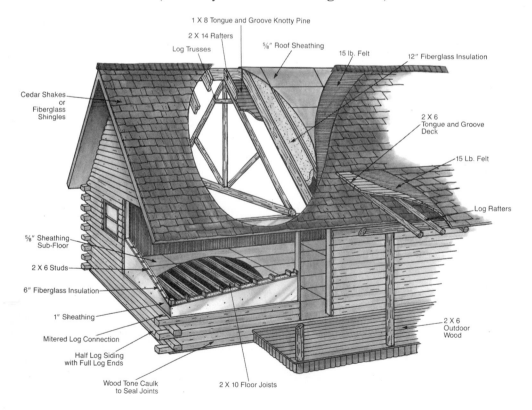

1 X 8 Tongue and Groove Knotty Pine
2 X 14 Rafters
⅝" Roof Sheathing
15 lb. Felt
12" Fiberglass Insulation
Log Trusses
Cedar Shakes or Fiberglass Shingles
2 X 6 Tongue and Groove Deck
15 Lb. Felt
Log Rafters
⅝" Sheathing Sub-Floor
2 X 6 Studs
6" Fiberglass Insulation
1" Sheathing
Mitered Log Connection
Half Log Siding with Full Log Ends
Wood Tone Caulk to Seal Joints
2 X 10 Floor Joists
2 X 6 Outdoor Wood

subcontractor is ready for the kit, you want to be sure it will be there. The same holds true even if the company you buy from does the kit construction.

WARRANTIES: Again, most companies offer limited warranties. Check to see what the warranty covers.

REFERENCES: Obtain the names of all people that have purchased homes from the companies you are interested in and call them. You can also find names of log home companies from advertisements, referrals, and the listing in the back of this book, reprinted with permission from *Log Home Living* magazine's *Annual Buyer's Guide*.

"Just the Logs, Please"

Keep in mind that log home companies are just that—log home companies. You should be buying just the logs and heavy timbers from them. Windows, doors, nominal lumber, and the like are always less expensive when purchased locally. Don't let yourself get talked into a total package when all you really want are the logs. The better log home companies will tell you this themselves.

A customized home with soaring ceilings and a breathtaking view.

CHAPTER 3

Buying the Land

If you don't already own the property for your log home, one of your first jobs as general contractor will be to select and purchase (or contract to buy) the land. (One should note, this is one of the many things that you would have to do even if you were employing a professional general contractor.) If you already own your land, you may still find information from this chapter useful. Buying land is not a difficult procedure. Although buying land is a very personal decision, there are certain guidelines to help you. One of the most important is how much you should spend. The second most important factor should be the location of the property. Is it in an area where there is good resale value? Most people don't consider selling their dream log home before building it, but resale should always be weighed. You never know what the future will bring and if you someday find yourself needing to sell your home, you will want to sell it quickly and for the best price. The lenders you will be talking to will also be concerned with resale value. A real estate broker can advise you about the resale value of a particular location.

How Much Should You Spend on Land?

How much you spend is something only you can determine, but a good real estate broker can help with location. A good real estate broker can also help with some of the other things we will be discussing. Let's look at how much you should spend.

Unless price is of no concern, a good rule of thumb on how much to spend on your land is approximately 15 to 20 percent of the appraised or market value of the finished house. That may seem difficult to determine since you don't know what the market value is yet, but it is not necessarily complicated. If you have an idea of the size log home you want, you can determine approximately how much it will cost to construct the house by figuring a rough estimate (covered in the chapter on estimating) or by getting an estimate from the manufacturer or his dealer. Then work backward to determine market value. Let's use the example in chapter 1 and the estimated cost of construction of $90,000. You can see that land costs based on construction costs are equal to 33⅓ percent, so you can simply multiply your rough estimate of $90,000 by 33⅓ percent. This comes out to $30,000, which is 18¾ percent of the market value.

As you can see, the first three phases of being a general contractor intermingle. You will have to know how much money you can spend in total (covered in the chapter on financing) and how much the house itself will cost (covered in the chapter on estimating) while you are still in the planning stage. If you have to spend more than what is suggested, plan for it in your overall budget. The size of the log home is the largest determining factor, other than the land, in affecting total costs. If you have to pay more for the land, you may have to build a smaller home. This may be one of the many compromises that you have to make throughout your job as general contractor. There is nothing wrong with making compromises. Make them carefully during all phases of construction, and you will find your role as general contractor a pleasant one.

Zoning, Restrictions, and Other Deciding Factors

When buying land, be sure that what you see is what you will be getting. The use of a real estate broker is strongly suggested. You or your broker will want to check for the following: zoning, restrictions, and utilities.

Zoning: This term is used to designate the use of a geographic area, as opposed to a single lot. It indicates where an area can be used for residential, industrial, commercial, or farm purposes. The zoning may combine some of the uses or denote other ones, depending on where you live. Zoning is determined by local government. It is meant to protect. You are usually assured that the use designated today will be the same tomorrow. If an area is zoned residential, it means that a gas station cannot be built on that property. However, zoning can be changed, which is another reason to use a knowledgeable broker. The broker should be able to tell by checking with planning commissions, zoning boards, etc., whether a certain area is likely to go through a zoning change in the future.

Restrictions: There are certain restraints placed on a particular lot or parcel of land by a previous owner or by the present owner. Restrictions are usually found in subdivisions, but they can exist on any property. Some restrictions are very strict, so you or your broker need to check for them. They are most often *recorded* (filed) with the *deed*. They could prevent you from parking a boat in your driveway, having a toolshed, or hanging your wash outside, or they might restrict your house as to size, style, or a number of other things.

Utilities: You or your broker will want to know what utilities are available for your land. Utilities include water, sewer, electricity, telephone, and even cable television. Water and sewer can be from a city system or community system. Find out how much it costs to tap in. If they are provided by a community system, be sure they are adequate and safe. You or your broker can check with the local health department, or you can hire a professional engineering firm to check the system. Such a firm is likely to be listed in the Yellow Pages under Engineers—Sanitary, Water, or some similar designation. If no water or sewer is available, or if one of the two isn't, you will need a well and/or septic system. Local health officials or a private firm will need to be consulted to determine if either a well or a septic system is feasible for the land. You need to check with the local power companies to see how much it will cost you for them to provide electricity or gas to the land. The same applies to telephone and cable television. All this checking does not take a great deal of time, and if you let a broker do it, it will take none of yours. Just be sure that all meets with your approval before you finalize the purchase of the land. This can be done by having in your contract to purchase the land a provision that allows you sufficient time to check on everything before finalizing (closing) the sale. This provision, or clause, is called a contingency. If things don't check out to your liking, be sure that this contingency allows you to get back any deposit (binder). The contingency should also have a clause that allows you time to find suitable financing. Always use a contract to purchase the land. Your broker or a real estate attorney can provide one.

Other factors in deciding on land could be:

1. Feasibility for having a solar home. Will you be able to face the house in the right direction in order to take advantage of solar energy? Will large trees be blocking the sun and therefore have to be removed?

2. Do you want a sloping lot? A sloping lot will require more foundation, and, therefore, a higher foundation cost. A sloping lot, however, lends itself more easily to having a basement that could have one or more walls open to daylight.

3. Do you want a lot with trees? Trees, especially large ones, enhance any home, and especially a log home, but clearing the lot for your home will cost more.

Hearthstone's picturesque Continental model captures the rustic charm found in log home living. Featuring a cathedral ceiling in the kitchen, this authentic, hand-hewn home offers three roomy bedrooms, two full baths, and a first-floor utility room for a total of more than 1,500 square feet of living space. Every Hearthstone home is hand-hewn by Appalachian craftsmen, using a foot adze.

CHAPTER 4

House Plans

Which comes first, the plans or the land, is not really that important except, as we saw in chapter 3, you may need the plans to determine how much you should spend on the land. You can make almost any house fit on almost any lot. But when you are ready to look for plans, your job has been made much easier for you by the log home companies. They offer an array of standard models that should please almost anyone's taste and budget. If not, most are willing and able to modify a standard model to your specifications or do custom plans at a price far below what you would have to pay an architect or home designer for conventional house plans. Some will subtract any charges from the price of the kit when you buy from them. They all will work with you to assure that you get the dream house you want.

How Big a House Do You Want?

Since the size of a home is the single most important factor affecting cost, it should be one of the first things you decide on. How large a home to build is based on your needs and budget. You can usually determine space needs based on previous dwellings in which you have lived and by visiting model homes. It is, of course, a subjective decision. What is large to one person may be small to someone else, and vice versa. If you are trying to keep overall size at a minimum, it is suggested that you keep the non–living areas as small as possible. These areas include bedrooms, baths, garages, and even kitchens. Instead of separate formal areas, you could have a large great room area. You could combine the kitchen with the dining area or den for a country-kitchen look, thus eliminating space, and so on.

You will find your log home company representatives very willing to assist you with your decision. They can also, as mentioned in chapter 3, give you a rough idea of what the finished house will cost, since they or their customers have built many. It can only be a rough idea because, as you will see, many other factors that go into finishing a house contribute to total cost and no two people will finish a house the same way or have the exact same labor or material costs. These and other factors and costs will be covered in chapter 7. A general rule of thumb to determine the total cost of the finished house, not including land, is

A sunburst window installed in the massive lodgepole pine wall affords views of the surrounding mountains. Maple flooring gleams against the hand-peeled logs handcrafted at Montana Log Homes.

two and a half to three and a half times the price of the kit. This will vary depending on what is included in the kit and local labor and other material costs. Ask the log home company what its formula is. You are lucky! You wouldn't have such a rule of thumb in conventional stick building!

What the Plans Include

You will find that reading plans is not difficult. You are interested in size. Anything structural on the plans will be understood by your subcontractors. Anything that is too difficult for them to understand can be explained to them by the log home company.

Your final plans, called blueprints, should include the following:

1. A foundation plan. Log homes can be built on any type of foundation, crawl space, basement, slab, pilings, or, in some areas, an all-weather wood foundation. The plan should indicate the complete foundation from the *footings* up to the *sill plate.* It won't, however, indicate the height of the foundation for a crawl space or basement, since that will vary with the slope of your land.
2. A floor plan for each level.
3. Exterior elevations for all four sides. Elevations are drawings showing what the outside of the building will look like.
4. A detail sheet. This shows a cross section of walls, roofs, vaulted areas, cabinets, and any part of the house that may not be clear from the floor plan. This sheet can vary in detail and content. Check with your log home company.
5. A specification sheet. This will list the materials that go into the building of your log home, right down to the carpet. An example, titled "Description of Materials," is at the end of this chapter.

Don't Plan Too Much Too Soon

An architect I know drew plans for a gorgeous home, perfect in every detail for his family and lifestyle. Only when the costs were figured did he realize that this was truly his "dream home" and would remain just that. He had to put these plans away and start over more realistically.

It is wiser to work backward, starting with the amount of money available through mortgages and cash, then subtracting the cost of the land. Buy your land first; the remaining amount is what you have to work with. Since size is the main factor of a home's cost, you can now get a more practical idea of how much house you can afford. Budget permitting, frills can be added from this point on.

To get more space for your money, go up, not out. Roofing, foundation, and heating are all more expensive when the house's size is increased horizontally rather than vertically.

Specifications (called specs) are a very important part of your plans. They list everything, including the logs, that will be going into your house. Included are the structural items as well as the decorative ones. The reason for having specifications is so that you, your lender, your suppliers, and your subcontractors know exactly what is going into your house. This is necessary for the purpose of controlling costs and quality. As you will see in chapter 8 on subcontractors, as well as in chapter 9 on suppliers, getting accurate and competitive *bids, contracts,* and *quotes* depends on accurate specifications. There are other forms available, and your log home company can help you complete any form you use. For some (or all) decorative items, you may not have made a selection, since specs come early in the planning process. In that case, you will need to budget a dollar amount for these items so that you can complete your cost estimate. For example, at the time the specs are completed, you most likely will not have selected carpet, wallpaper, stain colors, etc. But, by simply asking a supplier how much is normally to spent on an item, or asking your log home company, or relying on previous experience, you can come up with a reasonable figure. You will be talking to suppliers early in the planning process, and in chapter 9 you will see how to be assured that you are getting good advice and contractor's prices.

Form 2005
Form 26-1852
2/74

U. S. DEPARTMENT OF HOUSING AND URBAN DEVELOPMENT
FEDERAL HOUSING ADMINISTRATION

For accurate register of carbon copies, form
may be separated along above fold. Staple
completed sheets together in original order.

Form Approved
OMB No. 63–RO055

Proposed Construction

DESCRIPTION OF MATERIALS

No. _____
(To be inserted by FHA or VA)

Under Construction

Property address _____ City _____ State _____

Mortgagor or Sponsor _____   _____
(Name)   (Address)

Contractor or Builder _____   _____
(Name)   (Address)

INSTRUCTIONS

1. For additional information on how this form is to be submitted, number of copies, etc., see the instructions applicable to the FHA Application for Mortgage Insurance or VA Request for Determination of Reasonable Value, as the case may be.

2. Describe all materials and equipment to be used, whether or not shown on the drawings, by marking an X in each appropriate check-box and entering the information called for in each space. If space is inadequate, enter "See misc." and describe under item 27 or on an attached sheet. THE USE OF PAINT CONTAINING MORE THAN FIVE-TENTHS OF ONE PERCENT LEAD BY WEIGHT IS PROHIBITED.

3. Work not specifically described or shown will not be considered

unless required, then the minimum acceptable will be assumed. Work exceeding minimum requirements cannot be considered unless specifically described.

4. Include no alternates, "or equal" phrases, or contradictory items. (Consideration of a request for acceptance of substitute materials or equipment is not thereby precluded.)

5. Include signatures required at the end of this form.

6. The construction shall be completed in compliance with the related drawings and specifications, as amended during processing. The specifications include this Description of Materials and applicable Minimum Property Standards.

EXCAVATION:

Bearing soil, type _____

FOUNDATIONS:

Footings: concrete mix _____; strength psi _____ Reinforcing _____

Foundation wall: material _____   Reinforcing _____

Interior foundation wall: material _____ Party foundation wall _____

Columns: material and sizes _____ Piers: material and reinforcing _____

Girders: material and sizes _____ Sills: material _____

Basement entrance areaway _____ Window areaways _____

Waterproofing _____ Footing drains _____

Termite protection _____

Basementless space: ground cover _____; insulation _____; foundation vents _____

Special foundations _____

Additional information: _____

CHIMNEYS:

Material _____ Prefabricated (make and size) _____

Flue lining material _____ Heater flue size _____ Fireplace flue size _____

Vents (material and size): gas or oil heater _____; water heater _____

Additional information: _____

FIREPLACES:

Type: ☐ solid fuel; ☐ gas-burning; ☐ circulator (make and size) _____ Ash dump and clean-out _____

Fireplace: facing _____; lining _____; hearth _____; mantel _____

Additional information: _____

EXTERIOR WALLS:

Wood frame: wood grade, and species _____ ☐ Corner bracing. Building paper or felt _____

    Sheathing _____; thickness _____; width _____; ☐ solid; ☐ spaced _____" o. c.; ☐ diagonal; _____

    Siding _____; grade _____; type _____; size _____; exposure _____"; fastening _____

    Shingles _____; grade _____; type _____; size _____; exposure _____"; fastening _____

    Stucco _____; thickness _____"; Lath _____, weight _____ lb.

    Masonry veneer _____ Sills _____ Lintels _____ Base flashing _____

Masonry: ☐ solid ☐ faced ☐ stuccoed; total wall thickness _____"; facing thickness _____"; facing material _____

    Backup material _____; thickness _____"; bonding _____

    Door sills _____ Window sills _____ Lintels _____ Base flashing _____

    Interior surfaces: dampproofing, _____ coats of _____; furring _____

Additional information: _____

Exterior painting: material _____; number of coats _____

Gable wall construction: ☐ same as main walls; ☐ other construction _____

6. FLOOR FRAMING:

Joists: wood, grade, and species _____ ; other _____ ; bridging _____ ; anchors _____

Concrete slab: ☐ basement floor; ☐ first floor; ☐ ground supported; ☐ self-supporting; mix _____ ; thickness ____

    reinforcing _____ ; insulation _____ ; membrane _____

Fill under slab: material _____ ; thickness _____ ". Additional information: _____

7. SUBFLOORING: (Describe underflooring for special floors under item 21.)

Material: grade nd species _____ ; size _____ ; type _____

Laid: ☐ first floor; ☐ second floor; ☐ attic _____ sq. ft.; ☐ diagonal; ☐ right angles. Additional information: _____

8. FINISH FLOORING: (Wood only. Describe other finish flooring under item 21.)

LOCATION	ROOMS	GRADE	SPECIES	THICKNESS	WIDTH	BLDG. PAPER	FINISH
First floor ____							
Second floor ____							
Attic floor ____ sq. ft.							

Additional information: _____

9. PARTITION FRAMING:

Studs: wood, grade, and species _____ size and spacing _____ Other _____

Additional information: _____

10. CEILING FRAMING:

Joists: wood, grade, and species _____ Other _____ Bridging _____

Additional information: _____

11. ROOF FRAMING:

Rafters: wood, grade, and species _____ Roof trusses (see detail): grade and species _____

Additional information: _____

12. ROOFING:

Sheathing: wood, grade, and species _____ ; ☐ solid; ☐ spaced _____ "

Roofing _____ ; grade _____ ; size _____ ; type _____

Underlay _____ ; weight or thickness _____ ; size _____ ; fastening _____

Built-up roofing _____ ; number of plies _____ ; surfacing material _____

Flashing: material _____ ; gage or weight _____ ; ☐ gravel stops; ☐ snow gua

Additional information: _____

13. GUTTERS AND DOWNSPOUTS:

Gutters: material _____ ; gage or weight _____ ; size _____ ; shape _____

Downspouts: material _____ ; gage or weight _____ ; size _____ ; shape _____ ; number _____

Downspouts connected to: ☐ Storm sewer; ☐ sanitary sewer; ☐ dry-well. ☐ Splash blocks: material and size _____

Additional information: _____

14. LATH AND PLASTER

Lath ☐ walls, ☐ ceilings: material _____ ; weight or thickness _____ Plaster: coats _____ ; finish _____

Dry-wall ☐ walls, ☐ ceilings: material _____ ; thickness _____ ; finish _____

Joint treatment _____

15. DECORATING: (Paint, wallpaper, etc.)

ROOMS	WALL FINISH MATERIAL AND APPLICATION	CEILING FINISH MATERIAL AND APPLICATION
Kitchen ____		
Bath ____		
Other ____		

Additional information: _____

16. INTERIOR DOORS AND TRIM:

Doors: type _____ ; material _____ ; thickness _____

Door trim: type _____ ; material _____ Base: type _____ ; material _____ ; size _____

Finish: doors _____ ; trim _____

Other trim (item, type and location) _____

Additional information: _____

7. WINDOWS:

Windows: type _____ ; make _____ ; material _____ ; sash thickness _____

Glass: grade _____ ; ☐ sash weights; ☐ balances, type _____ ; head flashing _____

Trim: type _____ ; material _____ Paint _____ ; number coats _____

Weatherstripping: type _____ ; material _____ Storm sash, number _____

Screens: ☐ full; ☐ half; type _____ ; number _____ ; screen cloth material _____

Basement windows: type _____ ; material _____ ; screens, number _____ ; Storm sash, number _____

Special windows _____

Additional information: _____

8. ENTRANCES AND EXTERIOR DETAIL:

Main entrance door: material _____ ; width _____ ; thickness _____ ". Frame: material _____ ; thickness _____ "

Other entrance doors: material _____ ; width _____ ; thickness _____ ". Frame: material _____ ; thickness _____ "

Head flashing _____ Weatherstripping: type _____ ; saddles _____

Screen doors: thickness _____ "; number _____ ; screen cloth material _____ Storm doors: thickness _____ "; number _____

Combination storm and screen doors: thickness _____ "; number _____ ; screen cloth material _____

Shutters: ☐ hinged; ☐ fixed.   Railings _____ , Attic louvers _____

Exterior millwork: grade and species _____ Paint _____ ; number coats _____

Additional information: _____

9. CABINETS AND INTERIOR DETAIL:

Kitchen cabinets, wall units: material _____ ; lineal feet of shelves _____ ; shelf width _____

Base units: material _____ ; counter top _____ ; edging _____

Back and end splash _____ Finish of cabinets _____ ; number coats _____

Medicine cabinets: make _____ ; model _____

Other cabinets and built-in furniture _____

Additional information: _____

10. STAIRS:

Stair	Treads		Risers		Strings		Handrail		Balusters	
	Material	Thickness	Material	Thickness	Material	Size	Material	Size	Material	Size
Basement _____										
Main _____										
Attic _____										

Disappearing: make and model number _____

Additional information: _____

11. SPECIAL FLOORS AND WAINSCO

	Location	Material, Color, Border, Sizes, Gage, Etc.	Threshold Material	Wall Base Material	Underfloor Material
Floors	Kitchen _____				
	Bath _____				

	Location	Material, Color, Border, Cap. Sizes, Gage, Etc.	Height	Height Over Tub	Height in Showers (From Floor)
Wainscot	Bath _____				

Bathroom accessories: ☐ Recessed; material _____ ; number _____ ; ☐ Attached; material _____ ; number _____

Additional information: _____

22. PLUMBING:

Fixture	Number	Location	Make	Mfr's Fixture Identification No.	Size	Colo
Sink						
Lavatory						
Water closet						
Bathtub						
Shower over tub △						
Stall shower △						
Laundry trays						

△☐ Curtain rod △☐ Door ☐ Shower pan: material _____

Water supply: ☐ public; ☐ community system; ☐ individual (private) system.★

Sewage disposal: ☐ public; ☐ community system; ☐ individual (private) system.★

★ *Show and describe individual system in complete detail in separate drawings and specifications according to requirements.*

House drain (inside): ☐ cast iron; ☐ tile; ☐ other _____ House sewer (outside): ☐ cast iron; ☐ tile; ☐ other _____

Water piping: ☐ galvanized steel; ☐ copper tubing; ☐ other _____ Sill cocks, number _____

Domestic water heater: type _____; make and model _____; heating capacity _____

_____ gph. 100° rise. Storage tank: material _____; capacity _____ gal

Gas service: ☐ utility company; ☐ liq. pet. gas; ☐ other _____ Gas piping: ☐ cooking; ☐ house hea

Footing drains connected to: ☐ storm sewer; ☐ sanitary sewer; ☐ dry well. Sump pump; make and model _____

_____; capacity _____; discharges into _____

23. HEATING:

☐ Hot water. ☐ Steam. ☐ Vapor. ☐ One-pipe system. ☐ Two-pipe system.

☐ Radiators. ☐ Convectors. ☐ Baseboard radiation. Make and model _____

Radiant panel: ☐ floor; ☐ wall; ☐ ceiling. Panel coil: material _____

☐ Circulator. ☐ Return pump. Make and model _____ ; capacity _____

Boiler: make and model _____ Output _____ Btuh.; net rating _____

Additional information: _____

Warm air: ☐ Gravity. ☐ Forced. Type of system _____

Duct material: supply _____ ; return _____ Insulation _____ , thickness _____ ☐ Outside air in

Furnace: make and model _____ Input _____ Btuh.; output _____

Additional information: _____

☐ Space heater; ☐ floor furnace; ☐ wall heater. Input _____ Btuh.; output _____ Btuh.; number units _____

Make, model _____ Additional information: _____

Controls: make and types _____

Additional information: _____

Fuel: ☐ Coal; ☐ oil; ☐ gas; ☐ liq. pet. gas; ☐ electric; ☐ other _____ ; storage capacity _____

Additional information: _____

Firing equipment furnished separately: ☐ Gas burner, conversion type. ☐ Stoker: hopper feed ☐; bin feed ☐

Oil burner: ☐ pressure atomizing; ☐ vaporizing _____

Make and model _____ Control _____

Additional information: _____

Electric heating system: type _____ Input _____ watts; @ _____ volts; output _____

Additional information: _____

Ventilating equipment: attic fan, make and model _____ ; capacity _____

kitchen exhaust fan, make and model _____

Other heating, ventilating. or cooling equipment _____

24. ELECTRIC WIRING:

Service: ☐ overhead; ☐ underground. Panel: ☐ fuse box; ☐ circuit-breaker; make _____ AMP's _____ No. circuits __

Wiring: ☐ conduit; ☐ armored cable; ☐ nonmetallic cable; ☐ knob and tube; ☐ other _____

Special outlets: ☐ range; ☐ water heater; ☐ other _____

☐ Doorbell. ☐ Chimes. Push-button locations _____ Additional information: _____

LIGHTING FIXTURES:

Total number of fixtures _____   Total allowance for fixtures, typical installation, $ _____

Nontypical installation _____

Additional information: _____

INSULATION:

Location	Thickness	Material, Type, and Method of Installation	Vapor Barrier
Roof			
Ceiling			
Wall			
Floor			

HARDWARE: (make, material, and finish.) _____

_____

SPECIAL EQUIPMENT: (State material or make, model and quantity.   Include only equipment and appliances which are acceptable by local law, custom and applicable FHA standards.   Do not include items which, by established custom, are supplied by occupant and removed when he vacates premises or chattles prohibited by law from becoming realty.) _____

_____

_____

_____

_____

_____

_____

_____

MISCELLANEOUS: (Describe any main dwelling materials, equipment, or construction items not shown elsewhere; or use to provide additional information where the space provided was inadequate.   Always reference by item number to correspond to numbering used on this form.) _____

_____

_____

_____

_____

_____

_____

_____

_____

PORCHES:

_____

_____

TERRACES:

_____

_____

GARAGES:

_____

_____

WALKS AND DRIVEWAYS:

Driveway: width _____ ; base material _____ ; thickness _____ "; surfacing material _____ ; thickness _____ "

Front walk: width _____ ; material _____ ; thickness _____ ".   Service walk: width _____ ; material _____ ; thickness _____ "

Steps: material _____ ; treads _____ "; risers _____ ".   Cheek walls _____

OTHER ONSITE IMPROVEMENTS:

(Specify all exterior onsite improvements not described elsewhere, including items such as unusual grading, drainage structures, retaining walls, fence, railing and accessory structures.)

_____

_____

_____

LANDSCAPING, PLANTING, AND FINISH GRADING:

Topsoil_____" thick: ☐ front yard; ☐ side yards; ☐ rear yard to _____ feet behind main building.

Lawns *(seeded, sodded, or sprigged)*: ☐ front yard _____; ☐ side yards _____; ☐ rear yard_____.

Planting: ☐ as specified and shown on drawings; ☐ as follows:

_____ Shade trees, deciduous, _____" caliper.	_____ Evergreen trees. _____' to _____', B &
_____ Low flowering trees, deciduous, _____' to _____'	_____ Evergreen shrubs. _____' to _____', B &
_____ High-growing shrubs, deciduous, _____' to _____'	_____ Vines, 2-year _____
_____ Medium-growing shrubs, deciduous, _____' to _____'	_____
_____ Low-growing shrubs, deciduous, _____' to _____'	_____

IDENTIFICATION.—This exhibit shall be identified by the signature of the builder, or sponsor, and/or the proposed mortgagor if the latter known at the time of application.

Date_____   Signature _____

Signature _____

FHA Form 2005
VA Form 26-1852

PHASE II
THE FINANCING STAGE

This custom-designed home is manufactured with western red cedar logs by Mountaineer Log Homes. It features a circular dining room with plenty of glass for an elegant view.

CHAPTER 5

The Construction Loan

You need two types of loans to finance your log home. The first is the construction loan, and the second is the permanent loan, or mortgage. The mortgage comes into play after your home is built and is discussed in the next chapter. The construction loan allows you to pay your subcontractors and suppliers during the construction of your home.

How you obtain a construction loan is discussed in a moment, but first, here is how one works. After you apply for and receive a construction loan, the money is disbursed (given) to you in stages, called *draws*. These draws are in an amount equal to the percentage of completion of your home. For example, if at the end of the first month, your house is 25 percent complete, you will receive 25 percent of the amount of the construction loan. The percentage of completion is determined by the lender. Certain percentage points are given for certain items completed. A typical chart used by a lender to determine completion percentage is on the next page. The draws you receive are usually sufficient to cover expenses, and you usually receive the draw before you even get bills for those expenses. You may have to pay some expenses prior to receiving a draw, but they are not usually more than a few thousand dollars. You will get the money for these expenses back when you receive the next draw. If you don't have a few thousand dollars to cover these expenses, you could borrow enough from a commercial bank on an interim basis (a short-term note). Such an interim loan is not difficult to obtain, since the source of repayment will be your construction loan.

Paying for Your Kit

An interim loan is also a good way to pay for your log home kit, since it usually has to be paid for before you receive your first draw. However, some construction lenders will advance a draw for the purpose of paying for a kit. Be sure to check with your lender for its policy on this point.

The amount of money you can borrow for construction is usually the same as the permanent mortgage. When the house is completed, this mortgage will pay off the construction loan. In many cases, one lender can make both loans. The interest rate for a construction

Savings and Loan Association

Inspection Report and Disbursement Schedule

Date _____ Loan No._____

Borrower _____

Location: Street/Box #_____ on _____ Side of _____

between _____ and _____

in _____ Subdivision _____County

ID by _____

Date Construction to Begin _____

Contractor _____ Loan Officer _____

Date of Inspection

1. Start-up costs	1												
2. Rough clearing and grading	1												
3. Foundations	4												
4. Floor framing	4												
5. Wall framing	5												
6. Roof framing and sheeting	5												
7. Wall sheathing	1												
8. Roofing	2												
9. Well/water connection	2												
10. Septic tank/sewer tap	2												
11. Plumbing roughed	5												
12. Wiring roughed	3												
13. Heating-AC ducts	2												
14. Insulation	2												
15. Chimney/flue	2												
16. Siding/brick veneer	7												
17. Door frames set	2												
18. Windows set	3												
19. Particle board/flooring	2												
20. Inside walls	6												
21. Bath tile	2												
22. Outside trim	2												
23. Gutters	1												
24. Inside trim	3												
25. Doors hung	2												
26. Plumbing fixtures	4												
27. Cabinets	3												
28. Heat plant	2												
29. Exterior painting	2												
30. Interior painting	4												
31. Built-in appliances	2												
32. Electrical fixtures	2												
33. Carpet/floor finish	4												
34. Screens	1												
35. Drives and walks	3												
36. Cleaning	1												
37. Landscaping	1												
TOTAL	100												
Date													
INSPECTOR													
INSPECTOR													

(Courtesy of Rocky Mountain Log Homes)

Log homes and the outdoor lifestyle have always seemed to go together. This Rocky Mountain Log Home located near Missoula, Montana, is the "base camp" for this active family. Unique features include a standard concrete foundation/basement faced with half logs, vertical log accents to the top floor, and a standing seam metal roof.

loan is usually a little higher than that for a mortgage, but it is for a very short period of time. This interest is considered a cost of construction and is listed as such on the estimated cost sheet in chapter 7. You pay interest only on moneys received each month, not on the whole amount to be borrowed. Often construction loan interest can be paid out of the proceeds of the draws each month.

How to Get a Loan

If you follow the steps below, you should be able to obtain your construction loan acting as your own general contractor. You should have these steps completed before applying for your construction loan. Lenders will be more apt to make your loan if you do. You must show that you can do as good a job as general contractor as any professional can do. The lender's business is to lend money. Your job is to convince them they should lend it to you. If they see you are well prepared and eager to get started, you will be doing your job well.

1. Have your property either purchased or contracted to purchase with the contingency that you can obtain financing. If neither is possible at this point in time, then at least have your property selected.
2. Have your house plans and a *survey* of your property.
3. Have an accurate cost estimate complete (see chapter 7).
4. Have your major subcontractors lined up (see chapter 8).
5. Have your suppliers lined up with accounts opened, if possible (see chapter 9).
6. Have proof of your income for the last two years and a list of all debts and obligations.
7. If possible, have a letter of commitment for a mortgage. This will be discussed in the next chapter.
8. Have a positive attitude.

If one lender says no, try another. A lender's reluctance to make a construction loan to you acting as your own general contractor is usually based on fears that: 1. you will not be able to complete the project, leaving a partially finished home, and/or 2. the costs will run way over your estimate, making the home unaffordable to you. Since both of these things have occurred in the past, it is your job to convince them it won't happen in your case.

Kits Make It Easier

Using a log kit or shell should help alleviate those fears, because you can control costs more easily. Log homes are also easier to build and go up more quickly than most conventionally built houses. You can also show them pictures of finished homes, maybe even the one you are planning. That is something that's difficult to do with conventionally built houses. Be persistent but pleasant. Remember, and you can remind the lender, that hundreds of thousands of people like yourself have done it and will continue to do it. They got loans and you can, too.

Whether you are relaxing on the deck with a good book or eating at the dining room table, a log home offers an invitation to leisure. Even traditionally popular gathering spots such as the kitchen and dining area take on an added measure of warmth. The openness of design invites good times and get-togethers.

CHAPTER 6

The Permanent Loan

A mortgage is a loan that has the repayment stretched out over a very long period of time. Without mortgages there would be very few houses in this country, because few people could afford them. Mortgages, however, have changed a lot in the last few years. The most notable changes have been periodic shifts in the interest rate charged for a mortgage, with higher rates forcing many people out of the housing market. Another change is that there is now a wide variety of mortgages available. The variety exists to help *qualify* a potential buyer.

How Much Can You Borrow?

Qualifying simply means the lender thinks you can make the monthly payments on your home. Lenders use a variety of methods to make this decision. You can and should, as one of your first steps in planning your home, sit down and discuss mortgages with one or more lenders. They are the experts and their advice is free. They are some of the professional helpers mentioned earlier. You can discuss their requirements for qualification, types of mortgages and their respective interest rates, and which would be best for you. This would be merely a preliminary meeting. At this time you are not applying for the mortgage. You are only trying to find out how much you can borrow so you can determine how much house you can afford. Some real estate brokers call this prequalifying. Talk to more than one lender, because there are differences from lender to lender. The more you learn from them, the more comfortable you will feel later when you do apply. Even if they also make construction loans, it is best at this time not to discuss construction financing but rather to wait until you have followed the steps in chapter 5.

No Money Down

After it is determined how much you can borrow, you can then figure out how much house and land you can afford. For example (from chapter 1): If you qualify for a $120,000 mortgage, you should be able to afford the $160,000 house in the example. If you borrow the full $120,000, the cost of both the land and the house is covered by the mortgage. You will be in your home with no down payment. You will want to have your construction loan in the same amount. Your construction lender may want you to pay for the land before it makes the loan, but you will get that money back. Some construction lenders will make a land *draw* and pay off the land. Some construction lenders will allow the owner of the land to get the mortgage before they are paid. This is called a *lot subordination*. This arrangement would have to be worked out with an attorney and, of course, the landowner. You could also finance the land on a short-term basis with a commercial bank. If you need to use the land as *collateral,* this too would have to be handled as a lot subordination. Consulting with a real estate attorney will clarify all of this. Attorney's fees are a cost of construction, so don't worry about spending money for this advice.

Obviously, if you want to put cash into your home, you would lower the amount of mortgage money (and construction money) needed. When you borrow less, your monthly payments are lower, and you effectively beat interest rates. You also may put yourself into a better position for qualifying for a loan in the first place.

Letter of Commitment

If you are going to use one lender for construction financing and another for the permanent mortgage, you will need a *letter of commitment* from the permanent lender to present to the construction lender. This letter tells the construction lender that you have been approved for a mortgage. They then know that if they make a construction loan, it will be paid off. A letter

---

Appraisal Value

When you build a log home, its market value, or the amount it will sell for, is in most cases the same as that of a stick-built home of comparable size. The building costs, however, are higher for the log home. Buyers should be aware of this and realize that their budget may not afford them quite as large a home as planned.

Since mortgages are based on appraisal value, the owner/builder might not be able to borrow as much as he or she needs. Why? Log homes don't appraise according to the "true cost approach." Appraisals are based on comparable real estate sales, and there have not been enough recent log home resales to reflect a true cost approach.

---

of commitment can be obtained at any stage of your planning. You don't necessarily have to have selected land or plans yet. It is made to enable one to either look for an existing house to buy or, in your case, make arrangements to build. The letter will state that the finished house will have to pass final inspection by the lender. In the case of FHA/VA loans, the house will have to pass not only a final inspection, but also a series of inspections during construction. Again, you can discuss all this with a potential lender for further clarification.

PHASE III
THE ESTIMATING STAGE

Gastineau Log Homes specializes in oak log homes. Only the heartwood of selected varieties is used, to produce logs of exceptional strength and beautiful grain.

Featuring 6-by-12-inch timbers and dovetail construction, this log home was custom-designed to provide the best possible view of a lake and mountains.

CHAPTER 7

Cost Estimating

Estimating the cost of your log home will be one of your most important jobs as general contractor. Your estimate will determine what you can buy, and its accuracy will help you get your construction loan. Because you are building a log home, your job is made easier, because some of the hardest work is done for you by the log home company. In conventional construction, it is very difficult, if not impossible, to estimate all the materials that make up the exterior walls. How many bricks, how much siding, how many studs, or how much insulation, drywall, paint, trim, etc., go into a conventional wall is hard to guess. But in a log wall, it is all there. So are many of the other materials, such as the roof system, ceilings, and whatever else the company is providing. And you get a firm price on all of it.

Rough Estimating

For a quick, or rough, estimate, most log home companies have a rule of thumb they use to give you a general idea of the cost of their finished homes. The rule usually is two and a half to three and a half times the kit price. This rough estimate is, as its names implies, approximate. It should only be used as a guide in the initial stages of planning. Another way to obtain a rough estimate is to use the average cost of construction per square foot for your area of the country. Your log home company's local representative should be able to give you that figure.

Accurate Estimating

The only way to get an accurate estimate is to get estimates on as many of the separate items that go into a house as you can. Listed below is an estimate sheet. In getting these estimates, you will be contacting suppliers and subcontractors, opening accounts, and doing comparison shopping. When you have completed it, your role as general contractor will be almost over. You will have your plans, land, estimate, subcontractors, and suppliers all ready to begin construction. Following the list, each item is discussed.

Sample Estimate Sheet

ITEM	ESTIMATED COST	ACTUAL COST
1. Land		
2. Survey		
3. Plans and specifications		
4. Closing costs		
5. Insurance (fire)		
6. Construction loan interest		
7. Temporary utilities, permits		
8. Lot clearing and grading, lot staking, and plot plan		
9. Excavation (for a basement)		
10. Footings		
11. Foundation, fireplaces, and chimneys		
12. Foundation waterproofing, soil treatment		
13. Subfloor (if not in kit)		
14. Log kit		
15. Log kit freight		
16. Additional framing materials (if not in kit)		
17. Exterior trim (if not in kit)		
18. Windows and exterior doors (if not in kit)		
19. Kit construction labor (carpentry labor, including labor for subfloor and exterior trim)		
20. Roofing material (if not in kit)		
21. Roof labor		
22. Plumbing		
23. Heating, venting, and air-conditioning		

Sample Estimate Sheet (continued)

ITEM	ESTIMATED COST	ACTUAL COST
24. Electrical		
25. Concrete slabs		
26. Insulation (if not in kit)		
27. Water and sewer (or well and septic)		
28. Interior wall paneling or drywall—labor and materials		
29. Interior trim and doors		
30. Cabinets		
31. Interior trim labor (carpentry)		
32. Painting, staining, and preservative (if necessary)		
33. Appliances		
34. Light fixtures		
35. Floor covering		
36. Drives, walks, and patios		
37. Decks		
38. Cleaning and trash removal		
39. Wallpaper		
40. Hardware and accessories		
41. Landscaping		
42. Miscellaneous (unforeseen costs and cost overruns)		

With most of these items, you will be able to get accurate costs before you start building. With others you won't, but you can come reasonably close. As each item is completed during construction, you should enter its cost in the actual cost column next to the estimate. If the actual cost is more than the estimated cost, you can look for ways to lower costs in subsequent items. For example, you could use less expensive floor covering or appliances. You can even eliminate some, like wallpaper and garage doors. In this way, you have a reasonable amount of control over the total cost. Where to find each subcontractor is discussed in detail in chapter 8. Below is a discussion of the necessary items in your estimate.

1. **Land.** This is obviously an item that you will have a firm price on.

2. **Survey.** A survey will be required by your lender. Even if your lender did not require one, you would want a survey made of the property. A survey determines accurately the boundaries of the land you are buying. It should always be done by a registered surveyor. Cost will vary with the amount of land to survey, difficulty in locating corners and angles, and other variables, such as a surveyor's familiarity with the area. You can get a close estimate beforehand, however, over the telephone.

3. **Plans and specifications.** As with the land, you will have an accurate estimate of this item.

4. **Closing costs.** These costs can be explained and estimated by a lender, even before you apply for your loan. They can be obtained over the telephone. These costs vary with the amount of the loan. They can consist of service charge, *points,* attorney's fees for preparing the closing statements or documents and for certifying *clear title,* title insurance, prepaid fire insurance, preparing the *title,* taxes, *recording fees,* and any other fees the lender may charge. The total for most closing costs is usually in the neighborhood of 3 percent of the loan amount. If you can obtain both the construction loan and the permanent loan from the same lender, you can save money by avoiding duplication of some closing costs. Having two different lenders means having two sets of closing costs.

5. **Insurance.** You will be required by your lender to carry insurance on your home while it is under construction. This insurance is called a builder's risk policy, and it is necessary in the event of fire or damage. The extent of coverage and what exactly is covered vary with insurance companies. You should shop around by phone to get the most coverage for the least amount of money. A builder's risk policy does not cover people. At the advice of your insurance agent, you may want to obtain a general liability policy in case someone other than a subcontractor is injured on the job site. Your subcontrac-

(Courtesy of Rocky Mountain Log Homes)

The warmth of log walls and this antique tub will melt the stress away. Constructed by Rocky Mountain Log Homes, this unique home combines log walls with standard construction dormers for added texture and accents.

tors will have their own insurance coverage. It is very important that you be sure they provide you with a *certificate of insurance* proving that they do have insurance. A copy of a typical certificate of insurance is shown here. Your insurance agent can answer any questions you might have on insurance. You will have the exact cost for this item.

6. **Construction loan interest.** This amount can be estimated by a lender before you ever apply for the loan. The interest cost will vary with the size of the loan and the length of time it takes to complete your home. But a very close estimate can be obtained after you determine how much you are going to borrow.

7. **Temporary utilities, permits.** You will need electrical service, water, and possibly a portable toilet at your job site. Some local building codes require a toilet. A phone call

CERTIFICATE OF INSURANCE

ALLSTATE INSURANCE COMPANY   HOME OFFICE—NORTHBROOK, ILLINOIS

**Name and Address of Party to
Whom this Certificate is Issued**

Name and Address of Insured

INSURANCE
IN FORCE

TYPE OF INSURANCE AND HAZARDS	POLICY FORMS	LIMITS OF LIABILITY			POLICY NUMBER	EXPIRATION DATE
Workmen's Compensation **Employers' Liability**	STANDARD	STATUTORY * $ _____ PER ACCIDENT (Employer's Liability only) *Applies only in following state(s):				
Automobile Liability		**Bodily Injury**	**Each**	**Property Damage**		
☐ OWNED ONLY	☐ BASIC	$	PERSON			
☐ NON-OWNED ONLY	☐ COMPRE-HENSIVE	$	ACCIDENT	$		
☐ HIRED ONLY	☐ GARAGE	$	OCCURRENCE	$		
☐ OWNED, NON-OWNED AND HIRED	☐	**Bodily Injury and Property Damage** (Single Limit) $ _____ EACH ACCIDENT $ _____ EACH OCCURRENCE				
General Liability		**Bodily Injury**		**Property Damage**		
☐ PREMISES—O.L.&T.	☐ SCHEDULE	$	EACH PERSON			
☐ OPERATIONS—M.&C.		$	EACH ACCIDENT	$		
☐ ELEVATOR	☐ COMPRE-HENSIVE	$	EACH OCCURRENCE	$		
☐ PRODUCTS/ COMPLETED OPERATIONS		$	AGGREG. PROD. COMP. OPTNS.	$		
☐ PROTECTIVE (Independent Contractors)	☐ SPECIAL MULTI-PERIL		AGGREGATE OPERATIONS	$		
☐ Endorsed to cover contract between insured and	☐		AGGREGATE PROTECTIVE	$		
_____ _____ _____			AGGREGATE CONTRACTUAL	$		
dated_____		**Bodily Injury and Property Damage** (Single Limit) $ _____ EACH ACCIDENT $ _____ EACH OCCURRENCE $ _____ AGGREGATE				

The policies identified above by number are in force on the date indicated below. With respect to a number entered under policy number, the type of insurance shown at its left is in force, but only with respect to such of the hazards, and under such policy forms, for which an "X" is entered, subject, however, to all the terms of the policy having reference thereto. The limits of liability for such insurance are only as shown above. This Certificate of Insurance neither affirmatively nor negatively amends, extends, nor alters the coverage afforded by the policy or policies numbered in this Certificate.

In the event of reduction of coverage or cancellation of said policies, the Allstate Insurance Company will make all reasonable effort to send notice of such reduction or cancellation to the certificate holder at the address shown above.

THIS CERTIFICATE IS ISSUED AS A MATTER OF INFORMATION ONLY AND CONFERS NO RIGHTS UPON THE CERTIFICATE HOLDER.

Date_____, 19____   By_____

                                          Authorized Representative

U454-16
(6-75)

to the building inspection department will let you know. Portable toilet companies are listed in the Yellow Pages, and a quick call will give you the monthly rental charge. Even if you are not required to have one, it is recommended that you do. It will help avoid embarrassing moments. Electrical service is provided by your electrician. It consists of a temporary meter, *circuit breakers,* and receptacles, all mounted on a pole near the job site. This is called a *saw box,* and the electrical service is called *saw service.* The electrician usually provides and installs the saw box free if he is to wire your house. The monthly bill for electricity is, of course, your responsibility. The charge per month is very small, and your local power company can give you an estimate. Water is provided by paying your local utility department for service and having your plumber install a spigot at the water meter. If you are going to have a well in lieu of water service, you will have to have it installed before construction begins, unless you can borrow water nearby. Fees for water service can usually be obtained over the phone from your local utility. Costs for a well are discussed in number 27 below. A plumber usually will not charge for installing a spigot if he is to plumb your house. Permits that you will need and their costs can be obtained over the phone from your local building inspection department.

8. **Lot clearing and grading; lot staking and plot plans.** Your grading subcontractor can give you a contract price for this step after looking at your lot. He doesn't have to know exactly where the house is going to be positioned at this point. This step is discussed in chapter 10, but be sure the contract price includes hauling away stumps, debris, rocks, etc. Later he will need to know exactly where the house is to be positioned on your lot, and this is accomplished by placing stakes in the ground showing the outside corners of the house. This should be done by a registered surveyor/engineer, with your input, of course, as to where you want the house. The surveyor can give you a contract price. He may have to restake the house after the clearing and grading, because the stakes may get knocked out of place. Be sure you discuss any charge for coming back. Prior to staking, he will draw your *plot plan.* Be sure to get a quote for that also. After clearing, your surveyor can install the *batter boards.*

9. **Excavation.** If you are going to have a basement, your grading subcontractor can give you a contract price after looking at your lot. Again, he doesn't have to know exactly where the house is going to be positioned at this time. After he begins grading, you may need your surveyor to check the work in progress as to the proper depth and side clearances for foundation work and waterproofing (more on this in chapter 10). Be sure to get a price from the surveyor for this.

10. **Footings.** A contract price for *footings* can be obtained in advance from your footing

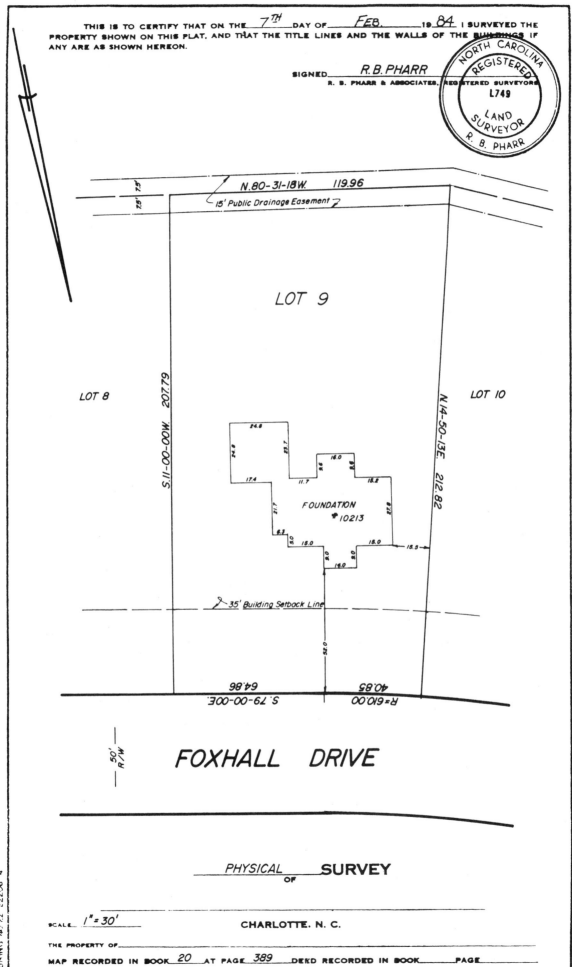

THIS IS TO CERTIFY THAT ON THE _7TH_ DAY OF _FEB._ 19_84_ I SURVEYED THE
PROPERTY SHOWN ON THIS PLAT, AND THAT THE TITLE LINES AND THE WALLS OF THE BUILDINGS IF
ANY ARE AS SHOWN HEREON.

SIGNED _R.B. PHARR_

R. B. PHARR & ASSOCIATES, REGISTERED SURVEYORS

NORTH CAROLINA
REGISTERED
L749
LAND
SURVEYOR
R. B. PHARR

N.80-31-18W.   119.96

15' Public Drainage Easement

7.5' 7.5'

LOT 9

LOT 8

S.11-00-00W. 207.79

N.14-50-13E. 212.82

LOT 10

24.8

24.8   23.7

16.0

17.4   9.6   9.6

11.7   18.2

21.7   27.8

FOUNDATION
#10213

6.3
5.0   18.0   18.0   18.5
8.0   8.0
19.0

35' Building Setback Line

52.0

64.86   40.85

S.79-00-00E.   R=610.00

FOXHALL   DRIVE

50'
R/W

PHYSICAL   **SURVEY**
OF

SCALE _1"=30'_   CHARLOTTE, N. C.

THE PROPERTY OF

MAP RECORDED IN BOOK _20_ AT PAGE _389_ DEED RECORDED IN BOOK   PAGE

subcontractor. He may want to give you an estimate, however, since it is difficult for him to estimate exactly how much concrete and/or labor it will take. This is all right if he will give you a maximum amount that he will not exceed. Footings are explained in chapter 10.

11. **Foundation, fireplaces, and chimneys.** Your brick and block mason subcontractor can give you only an estimate on these items. If he is good, he will come reasonably close. He will tell you how many bricks or blocks you will need and how much sand and mortar mix you will need to order, if he doesn't supply these items. He will tell you how much per brick and block the labor costs and your suppliers will tell you how much the materials cost. If fireplaces and/or chimneys are to be prefabricated, your supplier can give you a quote for the materials. Their installation is usually done by your carpenters, a sheet metal company, or your heating and air-conditioning subcontractor. All are capable of giving you a quote on the labor to install these items.

12. **Foundation waterproofing, soil treatment.** You can get quotes from waterproofing firms listed in the Yellow Pages. Soil treatment firms are also in the Yellow Pages.

13. **Subfloor.** If your log home company does not supply this, you can get a quote from a local lumber company. Often the log home company will give you a list of materials that go into the subfloor, but if not, the building supply company can make up such a list, called a *take-off,* from your plans and specifications. If you are building on a slab foundation, see number 25 below.

14. **Log kit.** You obviously will have this price. Again, be sure of what is included in the kit and how long the quoted price is good. Be sure the price is guaranteed long enough for you to complete all your planning, estimating, and financing arrangements.

15. **Log kit freight.** This amount can be calculated by the log home company. It will need to know where your building site is. If the site is such that large flatbed tractor trailers cannot get directly to it, be sure to allow some extra money for additional handling.

16. **Additional framing materials.** All that was said in number 13 above applies here. Again, you have to know exactly what is and what is not included in your kit.

17. **Exterior trim.** This consists of *fasciae, soffits,* moldings, *gable* trim, etc. Many log home companies include this in their kits because it affects the appearance of the finished product, and by supplying these materials, they are more assured of enhancing the final appearance of their product. If not supplied, all that was said in number 13 above applies here as well.

18. **Windows and exterior doors.** If these items are not supplied with the kit, a local build-

ing supplier can do a *take-off* and give you prices. An exact amount can be determined. Be sure that screens, *window grids, sash locks,* and storm windows and doors, if applicable, are included in prices. Prices can vary greatly by brand names, so shop carefully. Windows and doors are the greatest source of *heat loss* and *heat gain,* so insulated glass and/or storm glass is recommended and may be required by your local power company. A phone call to the power company will inform you on this. They may want to see your plans to calculate your requirements.

19. **Kit construction.** You can get a quote from your carpentry subcontractor based on the square footage of your home. The quote should be for all labor involved in building the subfloor, erecting the kit, drying in the house, and installing exterior and interior trim. Be sure to ask him if he charges extra for any items, and if so, these are to be explained to you, agreed on, and included in the quote. Extra labor costs might include cabinets, prefab fireplaces, decks, porches, sliding doors, insulation, paneling, extra moldings, stairs, *dormers,* roofing, and working at unusual heights. You should also understand that any changes or additions to the plans and specifications after you receive this quote will be an additional cost to you. Try not to make any, but if you do, obtain a revised quote.

20. **Roofing material.** If roof shingles are not included in your kit, any building supply company can do a *take-off.* They can also show you samples. Shingles vary greatly in price. A very close estimate can be obtained.

21. **Roof labor.** If your carpenters do not install shingles, and a good many don't, a roofing subcontractor can give you a contract price. Price is per square of roof area. A square equals 10 feet by 10 feet of area, or 100 square feet. Be sure that the contract includes any charge for *flashing,* installing *ridge vents,* and *capping.*

22. **Plumbing.** Plumbing contractors should include all labor and materials to plumb the house including the water heater and all other fixtures except appliances, such as dishwasher, disposal, and washing machine. It is important that your specifications be very clear as to plumbing fixtures. A trip to a plumbing supply company or two will be necessary to select the fixtures you want. The contract should also include any costs of connecting water and sewer lines to their source.

23. **Heating, venting, and air-conditioning (HVAC).** A firm quote is easily obtained for this item from your plans and specifications. Be sure it includes any venting for fans, stove vents, furnace venting for gas or oil, dryer vent, and any other required venting.

24. **Electrical.** From your plans and specifications, an electrical subcontractor can give you a firm quote. It should include all switches, receptacles, wires, panel boxes, circuit breakers, the wiring of all built-in appliances, heat/AC, any exterior lights, and possibly

security systems, intercom, and/or stereo wiring.

25. **Concrete slabs.** This item pertains to basement or garage concrete floors. An exact quote can be obtained for the concrete work from a concrete subcontractor. It should include any reinforcing wire, Styrofoam insulation, plastic film, and *expansion joints,* as required by code. Any required fill dirt or sand, and stone for drainage under the slab, should also be included in the quote.

26. **Insulation.** Insulation will be required for *subfloors, gable* ends, roofs, ceilings, and *dormer* walls. An exact quote can be obtained from an insulation subcontractor. If the insulation material is included with your kit, the labor to install can be contracted for with an insulation subcontractor, or with your carpenters.

27. **Water and sewer (or well and septic).** Your local utility company can give you tap-in fees. A septic system subcontractor, listed in the Yellow Pages, can give you an exact quote for your system. Local health officials, or private engineers, will determine its size and location on the property. Size, and sometimes location, affects price. A well driller should be able to give you an exact quote for a well. In some cases he may not be able to. Such circumstances could be rough terrain, rocky conditions, or his unfamiliarity with local water tables. If he has to charge a price per foot drilled, it is recommended that he give you a maximum amount that he cannot exceed. Be sure that his contract includes the size well, pump, storage and recovery tank, and/or any necessary filters. Be sure also that he guarantees quality water at a sufficient yield.

28. **Interior wall paneling or drywall (labor and materials).** From your plans and specifications, a building supply store can give you an estimate on materials for either paneling or drywall (Sheetrock). Your carpenters can give you an exact quote on installing paneling, and a drywall subcontractor can give you an estimate on installing and finishing

---

Unequal Pricing for Same Product

The two most expensive items in a log house are the windows and the logs. Windows are 90 percent glass but can vary in price by as much as 400 percent. There is no logical reason for this; glass is glass. Shop carefully and don't let brand names sway your sound judgment and ultimate decision.

This also holds true for each type and thickness of wood. As long as you are comparing 6-inch-diameter cedar logs with 6-inch-diameter cedar logs, or pine logs with pine logs of the same size, avoid worrying about brand names and opt for the best value instead. Just worry about the bottom line.

(Courtesy of Brentwood Log Homes)

The Merrylog is a model offered by Brentwood Log Homes.

drywall. Some drywall subcontractors also supply the drywall. Be sure their estimate includes taping joints, finishing with at least two coats of filler (called mud), sanding, and hauling away scraps, if possible.

29. **Interior trim and doors.** You will find a wide variety of styles and prices for these items. You will need to visit a building supply store to select them. They can do a *take-off* after you make your selection and give you an estimate. Interior trim would include moldings, stairs, handrails, and shelving. Be sure to include interior trim selections in your specifications.

30. **Cabinets.** These include kitchen cabinets, bath vanities, and possibly bookcases. An exact amount for all can be obtained from your plans by a building supply company or a cabinetmaker, who can be found in the Yellow Pages or by word of mouth. If the labor to install the cabinets is not included in the quote, get a quote from your carpenters.

31. **Interior trim labor.** Your carpenter can give you an exact quote based on your plans and specifications. It is usually based on a dollar amount per square foot. Be sure it includes the installation of all the trim you selected, the cabinets, vanities, and bookcases if necessary. As in number 19, Kit construction labor, you don't want to be charged for any extras at the end.

32. **Painting, staining, and preservative.** You can obtain an exact quote for painting and staining, as well as for log preservation if necessary. Do not pay by the hour! Your quote should include labor and materials. As per the advice of your log home company, you may or may not have to have your logs treated with a preservative. If you do, they can recommend a company that should be able to give you an exact quote.

33. **Appliances.** Appliances to include in the estimate are those considered built-ins. These would include dishwasher, range, ovens, and disposal. Prices vary considerably by manufacturer and model. You should shop more than one supplier, unless you know what you want from previous experience. An exact amount for your selections is easily obtained.

34. **Light fixtures.** This item usually includes floodlights, any decorative lighting, indirect lighting, doorbells, intercoms, and security systems. Most lighting supply companies have salespeople who can help you plan your needs and give you an exact quote. There is no charge for this service.

35. **Floor covering.** All these estimates are arrived at by measuring the square footage based on your plans. A floor covering supplier can give you an accurate quote for carpet or vinyl after you have made a selection. Price and quality vary widely. If you don't wish to decide at this point, he can give you an idea how much you might want to use as an allowance figure in your estimate. For wood floors, you can obtain a close estimate by getting an estimate on the wood from a building supply company, a quote for installing from your carpenter, and a quote on sanding and finishing from a floor finisher. For slate, tile, stone, or brick floors, an exact quote can be obtained from a tile subcontractor found in the Yellow Pages.

36. **Drives, walks, and patios.** Depending on what these items are to be made of—concrete, asphalt, or crushed stone—an estimate can be made by a subcontractor from a copy of your *plot plan,* which is the survey showing where the house will be located on the lot. Your input will be required as to widths, size, length, and materials used. One subcontractor may be able to give you a quote using any material. Quotes should, of course, include all labor and materials.

37. **Decks.** Decks should be shown on your plans, and a building supply company can do a

(Courtesy of Greatwood Log Homes)

Tucked away on a country lane is Moorland, a three-bedroom, three-full-bath designer series home from Greatwood Log Homes. Featured are large rooms, a country kitchen, a large master suite with its own master bath, a great room, a den, and a two-car garage. Total living area is 2,830 square feet. Premier log styles feature big cedar logs with large 9- to 10-inch diameters and random timber lengths ranging from 10 to 20 feet.

take-off as to materials. As mentioned in number 19 above, decks are usually an extra with your carpenter. He can, however, give you an exact quote.

38. **Cleaning and trash removal.** If you have never had a house built before, you might wonder why you have to clean a new house. The amount of trash generated will amaze you. However, any of the professional cleaning service companies in the Yellow Pages can give you an exact quote for cleaning, even from your plans. Their cleaning should include windows, both sides, bathrooms, cabinets, and everything inside the house. These companies, in most cases, include trash removal in their quote. If not, your landscape

subcontractor or your grading subcontractor from number 8 above can give you an estimate based on previous experience.

39. **Wallpaper.** Any wallpaper store can give you an estimate of what an average allowance might be from your specifications, since you probably won't be selecting your wallpaper just yet. When the time comes, they can assist you in selection and help in determining the amount of paper necessary.

40. **Hardware and accessories.** This will include doorknobs and locks, doorstops, towel bars, and mirrors. A building supply company can give you an estimate from your plans of what an average allowance would be. Prices vary considerably, and you can spend anywhere from a few hundred dollars to thousands on this item.

41. **Landscaping.** After looking at your land, seeing where the house will sit, and discussing what you want or need, a good landscape subcontractor can give you an exact quote.

42. **Miscellaneous.** It is impossible to plan for every cost in construction, so allow for the unknown. Take the total of the first forty-one items and multiply by 5 percent. This should cover not only unforeseen costs, but most cost overruns as well. In addition to this, factor in any items not included above, such as garage doors, stonework, swimming pools, and *chinking,* that may apply to your home.

All Costs Included

At the estimation stage, certain costs that are inherent to log homes are often overlooked. If you know about these costs and plan for them, you can avoid unpleasant surprises at the end of your construction project.

For example, caulking a log home can often run to several thousand dollars and is frequently missed in the cost estimating phase of planning. Oops! There goes the whirlpool you wanted. Other expenses often overlooked include stain, wood preservative, crane service, extra labor, and delivery. Any one of these might be enough to blow even the most carefully planned building budget.

PHASE IV
THE BUILDING STAGE

Nestled against the Tetons near Jackson Hole, Wyoming—big snow country—this Rocky Mountain Log Home is draft-tight and energy efficient. This home uses dried lodgepole pine, a Swedish cope-joining method, and a unique spline-and-gasket system to ensure weather-tight efficiency for the life of the home.

CHAPTER 8

Subcontractors

In this chapter we will discuss how to find good subcontractors, how to contract with them, how to schedule them, how to work with them, how and when to pay them, and how to inspect their work.

What Is a Good Subcontractor?

A good subcontractor does quality work at a reasonable price and is reliable. Determining quality is somewhat subjective. What one person considers good work, another may not. There is no perfection in construction, but as long as you and your team of inspectors, whom we will discuss below, are satisfied, that is all that should matter. If a subcontractor does quality work for one general contractor, he usually does so for another, since he depends on good references for his livelihood. This points out the importance of getting and checking references. It is recommended that you get at least three references for each subcontractor you are planning to hire. If any subcontractor refuses to give references, find someone else. In checking references for quality, you can also check on a subcontractor's price and reliability. Reliability can be as important as quality. If a subcontractor does not show up when scheduled, he can delay your entire job. This costs you money in construction interest.

Finding Good Subcontractors

There are several ways to find good subcontractors. Here is a list of most, if not all, of the subcontractors you will need and how to find them. NOTE: In some parts of the country, subcontractors may be called by different names.

1. **Real estate broker.** You can find a broker from brokers' ads in the newspaper, from word-of-mouth referral, or from the Yellow Pages. You may need one to purchase your land.

2. **Attorney.** Find on the referral of friends, real estate firms, or lending institutions or from the Yellow Pages. A real estate specialist will work faster for you.

3. **Lending officers.** These people will be at the banks that will be loaning you money, for both construction financing and permanent financing. No references are necessary here.

4. **Insurance.** You can use the insurance agent you now use for car insurance, renters' or homeowners' insurance, etc.

5. **Surveyor.** You can find through the Yellow Pages or on referral from real estate firms, attorneys, or lenders. Be sure the one you choose is licensed (registered) and insured.

6. **Log home company.** *Log Home Living* magazine's *Annual Buyer's Guide* is a good source for locating the log home company that is right for you. Contact Home Buyer Publications Inc., 610 Herndon Parkway, Suite 500, Herndon, VA 22070, (703) 478–0435 or (800) 826–3893. You can also find names of log home companies from advertisements, referrals, and the listing in the back of this book.

7. **Carpenter.** This is one of the most, if not *the* most important of your subcontractors. It is recommended that you line him up early in the planning process. He is also a good source for finding other subcontractors that you will need, since he works with most of them every day and knows both the quality of their work and their reliability. It is wise to get a carpenter who has erected a log home before and, preferably, one of the kind

---

Inexperienced Carpenters

Building a log home is not the same as building a stick-built house. Different carpentry skills are required. Many expensive mistakes can be avoided if the carpenters chosen for the job have completed at least one log home construction project. Carpentry tricks of the trade are learned on the job, not from books or manuals. Don't let your house be the first experience your carpentry crew has with building a log home.

Examples of possible on-the-job training errors include:
- trimming off roof rafter overhangs,
- improperly cutting logs for window and door openings,
- placing or stacking logs incorrectly,
- marring the log surfaces by using force behind the hammer rather than skill,
- leaving out required insulation between logs.

You can easily avoid all of these mistakes by finding out if your carpenters have any prior experience with log homes.

(Courtesy of Real Log Homes)

A very popular Real Log Homes design, the Cavendish is a cozy 985-square-foot two-bedroom log home with an exposed-beam cathedral ceiling in the living room.

you are buying. The log home company's representatives often have names of carpenters familiar with their product. If the company is to erect the shell, you will still need a good carpenter for trim, decks, etc. Other ways to find this subcontractor are from a building supply company or real estate firm, on referral from friends, or by stopping by a job under construction. You could also call a general contractor whose homes you have seen and whose workmanship you have admired. It's done all the time. Many general contractors don't build enough homes in a year to keep their carpenters busy full-time. They shouldn't mind giving you the name of their carpenter, since you are building only one house and will not seriously affect their scheduling of future jobs.

8. **Grading and excavation subcontractor.** Found through referrals from friends, your carpenter, real estate brokers, the Yellow Pages, or sand and gravel suppliers listed in the Yellow Pages; from a job under construction; or from a professional general contractor.

9. **Footing subcontractor.** This subcontractor is often the same as number 8, but if not, you can use the same sources as outlined in number 8. In different parts of the country, different types of footings are used and different names are applied to the process of this beginning part of your structure; for example, in some places pilings are used in lieu of footings.

10. **Brick and block masonry subcontractor.** This is the subcontractor who will build your foundation. Carpenters often know good brick masons. Also, referrals from friends, brick suppliers, professional general contractors, the Yellow Pages, or stopping by a job in progress are ways of finding this subcontractor.

11. **Waterproofing subcontractor.** For waterproofing your foundation, it is recommended that a professional company listed in the Yellow Pages under Waterproofing or another such heading be used to do your work. Almost anyone can do the job, but they may not guarantee their work. Water problems can be a bother, so it is best to have a pro take care of them early.

12. **Roofing subcontractor.** Use the same sources as number 7 and the Yellow Pages.

13. **Plumbing subcontractor.** You can get referrals from plumbing suppliers, friends, your carpenter, a job under construction, the Yellow Pages, or general contractors.

14. **Heating, venting, and air-conditioning (HVAC).** Same as number 13, and the Yellow Pages.

15. **Electrical subcontractor.** Same as number 13 and from an electrical supply company.

16. **Concrete (finisher) subcontractor.** Same as number 7.

17. **Insulation subcontractor.** From the Yellow Pages.

18. **Drywall subcontractor.** Same as number 7.

19. **Painter.** Same as number 7.

20. **Flooring subcontractor(s).** Same as number 7 and in the Yellow Pages. This subcontractor is usually a supplier.

21. **Cleaning service.** Same as number 7 and in the Yellow Pages.

22. **Wallpaper hanger.** A wallpaper store is usually a good source for this subcontractor, as are the sources listed in number 7.

23. **Landscaper.** Same as number 7 and in the Yellow Pages.

Subcontractor Bids and Contracts

Subcontractors contract with you to perform a certain task at an agreed-upon price (quote or bid). Because of this arrangement, they are not considered your employees by the government, so you needn't worry about withholding taxes. You will need to file Form 1099 with the Internal Revenue Service on all subcontractors that aren't corporations. The form merely shows how much money you paid them.

A sample subcontractor's contract is at the end of this chapter. It is merely an example. You should have your attorney draw one up for your use. You will notice that the contract provides spaces for the quote, work to be performed, insurance information, and the terms of payment. You should get three or four quotes or bids, based on your contract form, from each type of subcontractor. To ensure that you are comparing "apples with apples" when getting quotes, be sure your form is followed explicitly. Also, never pay a subcontractor by the hour rather than by the job. It is too easy to go over your budgeted allowances with subcontractors who are paid by the hour. Make sure that your specifications clearly state the materials each subcontractor is to provide if those materials are in his contract. By getting three or four bids, you can also get a better feel for costs in your area.

Scheduling Subcontractors

Scheduling subcontractors is not difficult and should require little of your time. In most cases they will be scheduled in accordance with the sequence of building steps outlined in chapter 10. After you have accepted a subcontractor's bid, you can let him know when you will need him, based on where he fits in the sequence and when you plan to start. There is an estimate of time for each step, so you should be able to give him a rough idea. Most scheduling is done by phone in the evening or on weekends, so your time involved is minimal. After construction begins, you can schedule better. Some of your subcontractors might keep an eye on job progress for you, in order to schedule themselves. You can ask if they would be willing to do this. If one subcontractor can't fit into your schedule after you start, then you will have to decide whether to wait for him or find another to do the job for you. If it is going to be a long wait, more than a few weeks, you might be better off finding another. Your decision will also depend on how much of your house has been completed. In the beginning, the construction interest that you are paying is minimal, but later in the process of building, it is higher. You can't afford to wait too long when the house is nearly completed and your interest payments are higher.

(Courtesy of Amerlink Log Homes)

Here is a fine example of how log interiors adapt to contemporary lifestyles. Amerlink has a new line of contemporary homes featuring the look of beveled siding, which blends beautifully into any residential setting while maintaining the natural beauty and benefits of solid wood construction.

Working with Your Subcontractors

If you have done your homework and have checked references carefully, the best way to work with your subcontractors is to leave them alone. Let them perform their jobs. They are the pros. If they have a question or need something, have them call you. You might want to install a job phone. It is inexpensive and will save you and your subcontractor's time. It can be locked up at night or taken home.

Paying Your Subcontractors

In the sample contract at the end of this chapter, you will notice the space provided for terms of payment. Terms of payment vary for different subcontractors and for different locales. But one good rule to follow is: Never pay for anything in advance! Pay only for work completed. However, some subcontractors will need to be paid on a weekly basis, and some when a certain part of their total job is complete. If a subcontractor needs a weekly payment, called a *draw,* then you will have to determine what percentage of the contracted job has been completed. It really is not difficult; let common sense prevail.

The subcontractors that might need a draw usually are brick masons, carpenters, and painters. If, for example, your carpenter is halfway through trimming out your house (he has installed one-half of the materials you ordered for trim), you can give him up to one-half of his contract price for trimming. Actually, a little less would be better, and 10 percent less than one-half is the norm. If your brick mason has laid a certain number of bricks or blocks, that is what you pay him for, less 10 percent. The amount of draw to release to your painters is more of a guesstimate, but be careful not to overpay.

Subcontractors that are paid for a part of their job are electrical, plumbing, and HVAC subcontractors. When their *rough-ins* are complete and have been inspected, you will pay them for a pre-agreed-upon percentage of their total quote. Be sure that this amount is in their contract. All other subcontractors are paid after they have completed their work and that work has been inspected. Some subcontractors (other than the ones requiring draws) can wait until you get your construction draw (thirty days or less) before getting paid, and this is to your advantage. Ask when discussing terms of payment in the contract.

Inspecting a Subcontractor's Work

In most areas there are building inspectors that inspect the critical stages of construction for quality and compliance with codes. If not, you can hire private professionals to inspect the work on your home. Other less critical stages of construction, such as painting, can be inspected by you, using your good common sense. Payments due to subcontractors should be made only after inspection of the work they have done. Inspections that are usually performed by your local building inspection department or professional inspector might be:

1. Temporary electrical service *(saw service).*
2. *Footings,* done before the concrete is poured.
3. Foundation.

Shown above is an example of Gastineau Log Homes' oak half-log construction with full-log corners. Gastineau offers a choice of wood species: oak, walnut, cedar, or pine. The exterior and interior profiles can be round or flat. Either full- or half-log construction is available.

4. Well and septic systems.
5. Concrete slabs, done before the concrete is poured.
6. Electrical, plumbing, HVAC *rough-ins*.
7. Structure, called framing inspection in conventional building. This is performed after the subcontractors in number 6 are finished, to check for structural soundness.
8. Insulation.
9. Final inspections for electrical, plumbing, HVAC, and the structure again to be sure that all codes are complied with and that everything works and is safe. In areas where there are building inspection departments, you probably will have to have completed number nine before you can get permanent electrical or gas service. Your lender may require the same inspections before permanent financing is given.

Other inspections may be required in your area. They are for your protection. If at any time you are in doubt as to whether or not a job is done properly, call your building inspector or hire a professional inspector and find out. Professional inspectors may be called home inspection firms or engineering firms. In some areas you will find home inspection firms listed in the Yellow Pages. They can usually perform any necessary inspections. In other areas firms may be listed in the Yellow Pages under headings for the specific inspections they perform. Most firms are listed under the heading "Engineers" and followed by their specialty, i.e., foundations, HVAC, consulting. You can also hire an architect or a professional general contractor for any inspections.

Don't worry about spending money for inspections, even if it is in the hundreds of dollars. You are saving thousands by being your own general contractor, and these inspections will assure you that your home meets the codes and standards that both you and the building inspection department set for it.

A cozy log cabin interior by Maple Island Log Homes.

Carpentry Labor Contract

TO: _____   SUB CONT.: _____
     (YOUR NAME)

_____
     (ADDRESS)

_____

DATE: _____   JOB ADDRESS: _____

     _____

OWNER: _____   _____

AREA: Heated _____ sq. ft.
      Unheated _____ sq. ft.
      Decks _____ sq. ft.

CHARGES

Kit Construction @_____ sq. ft. x _____ sq. ft. = $_____
Exterior Trim @_____ sq. ft. x _____ sq. ft. = $_____
Interior Trim @_____ sq. ft. x _____ sq. ft. = $_____
Decks @_____ sq. ft. x _____ sq. ft. = $_____
Setting Fireplace $_____
Setting Cabinets $_____
Paneling $_____
Misc. $_____
          TOTAL CHARGES $_____

Terms of Payment: _____

_____

Insurance Information: _____
     (NAME OF INSURED)
INSURANCE COMPANY: _____ POLICY NUMBER: _____

SIGNED: _____ DATE:_____
     (YOUR NAME)
SIGNED: _____ DATE:_____
     (SUBCONTRACTOR)

Subcontractor's Invoice

TO: _____

DATE: _____

Request Number: _____

CONTRACTOR: _____

_____

_____

CONTRACT NUMBER: _____

CHANGE ORDER NUMBER: _____

WORKMEN'S COMP. INS. CO.: _____

JOB NAME	JOB NO.	DESCRIPTION OF WORK	AMOUNT

Work Completed in Accordance with Contract:

_____

(CONTRACTOR)

Total: _____

LESS RETAINER: _____

_____

_____

_____

NET AMOUNT DUE: _____

CHAPTER 9

Suppliers

Other than your log home company, you probably will be buying from some or all of the following suppliers:

1. Sand and gravel company. For sand for your brick masons, dirt for *backfilling* and landscaping, gravel for drives, etc. Some subcontractors supply these items, so you may have no need for this supplier.
2. Block supplier. For foundation block.
3. Brick supplier. For face (decorative brick for foundations, chimneys, etc.).
4. Concrete supplier. Many concrete subcontractors furnish concrete in their contract price.
5. Building supply company. For any framing materials not included in your kit and/or doors and windows, etc., if not included. Also for interior trim and many of the items carried by the suppliers listed below.
6. Floor covering supplier. For flooring needs, and usually for countertops as well.
7. Light fixture supplier or electrical supply company.
8. Paint store. To select colors only. Paint is usually supplied by the painter in his contract price. Paint stores also carry wallpaper.
9. Appliance store.
10. Tile company. For tile, slate, marble, and decorative stone.
11. Specialty stores for items such as solar energy systems and fences.

Opening Accounts and Getting Discounts

To buy supplies at a builder's discount, you will need to open a builder's account with each supplier. This is not difficult. All you have to do is explain to the management that you are building a house and that you would like to receive a builder's discount. Simple as that! They now know you are not just a weekend do-it-yourselfer buying in small quantities. They also know that many of their professional accounts started out just like this. Builders' discounts vary and in some cases are very small. But it all adds up. So does the sales tax, and it is completely tax deductible, another savings for you. To open an account with a supplier, you most likely will need at least three credit references, such as Sears, and a bank reference. They may also want you to inform them of your construction lender when you get one. Their terms of payment are quite different from Sears, as you will see.

Paying Your Suppliers

Most suppliers who deal with professional general contractors on a daily basis have their terms of payment set up to aid the general contractor. Payment for supplies purchased in one month is not due until the following month. This allows the general contractor (you) to have time to get a construction *draw*. For example, if you purchased brick on June 1, it may not have to be paid for until July 30. Usually a 2 percent discount is given if you pay by the tenth of the following month, July 10 in this case. Always ask if a discount is given, because often it is not stated on the invoice or monthly statement.

Bookkeeping

Since you are only building one house, a checkbook should provide sufficient bookkeeping records. You can transfer amounts from your check register onto the actual cost column of the estimate sheet. You should, of course, open a separate checking account to be used solely for payment of construction bills. If you want to use a more refined means of record keeping, such as a home computer or simple accounting forms from an office supply store, fine. Be sure you keep a record of any sales taxes paid, because you can claim these taxes on your personal income tax.

No Deposit

In building a log home, as in any construction project, deposits are dangerous and should be avoided if possible. When you pay money in advance, you have no guarantee that the company will be in business the next day, no matter how large that company is. By providing a company with a deposit, you are acting as its bank, financing its day-to-day activities.

It is preferable to have your lender guarantee payment on delivery. Some sort of deposit is probably unavoidable, but keep it as small as possible, certainly not more than 10 percent of the total cost of the item being ordered.

CHAPTER 10

Building Your Home

When you reach this point, your work as a general contractor is almost complete. Now your team of professionals can go to work and build your dream log home. In this chapter the sequence in which your home will be built is discussed, and the average time for each step is indicated. Actual time to complete any step will vary due to factors such as weather, techniques of construction, and other reasons.

Steps of Construction

Here is a list of the steps of construction. Each step is discussed following the list.

1. Ordering the kit or shell and getting permits. 1–3 hours.
2. Staking the lot and house. 1 day.
3. Clearing and excavation. 1 day to 1 week.
4. Ordering utilities, temporary electrical service, portable toilet; getting insurance. 1–3 hours.
5. Footings. 1–2 days.
6. Foundation, waterproofing, and soil treatment. Call for a foundation survey when complete. 1 day to 2 weeks.
7. Plumbing rough-in, if slab foundation. 2–4 days.
8. Slabs. 1–3 days.
9. Kit construction and drying-in, exterior trim. 1–3 weeks.
10. Chimney and fireplace(s). 1 week.
11. Roofing. 1–3 days.
12. Plumbing, HVAC, and electrical rough-ins. 2 weeks.
13. Insulation. 2 days.
14. Hardwood flooring and carpet underlayment. 2–5 days.
15. Drywall or paneling. 2 weeks.
16. Interior trim. 1–3 weeks.
17. Painting and staining. 2–3 weeks.

18. Tile, countertops, etc. 1–2 weeks.
19. Trim out plumbing. 2–4 days.
20. Trim out HVAC. 1–2 days.
21. Trim out electrical. 2–4 days.
22. Floor finish and/or carpet. 2–5 days.
23. Cleanup. 2–3 days.
24. Drives and walks. 2–4 days.
25. Landscaping. 1–3 days.
26. Final inspections. 1–2 days.
27. Loan closing. 1 hour.
28. Enjoyment. A lifetime!

The Steps Explained

1. **Ordering the kit or shell.** Your log home company should inform you how far in advance you need to order your kit or shell. When your foundation is completed, you will want to have your kit there or on its way. By reading this chapter and discussing schedules with your subcontractors, you can approximate reasonably well when this time will be. If something delays that time, shipping can be delayed. Most companies will work closely with you on delivery dates.

    At this time you should obtain any necessary permits. A call to your building inspection department and/or health department will inform you of the necessary permits and the procedure to follow in obtaining them. It can most often be done on a lunch hour or two.

    You also can arrange for your builder's risk insurance at this time. It needs to be in force prior to starting construction.

2. **Staking the lot and house and installing batter boards.** Your surveyor should check to see that all boundary stakes are accurate. They could have been moved since you purchased your land. Then you and he can determine precisely where the house will "sit" on the lot, and you will place stakes showing all the corners of the house. This can be done on your lunch hour, after work, or on the weekend. Your surveyor will make sure that your house is not in violation of any setback restrictions. The corner stakes will let your clearing and excavation subcontractor know where to clear and/or excavate. Usually the surveyor will also place offset stakes indicating the corners far enough away so that they won't be disturbed, and the actual corners can be relocated if need be. Your

(Courtesy of Greatwood Log Homes, Inc.)

By utilizing Greatwood's professional staff of design engineers and consultants, you can customize any of their existing floor plans or create something entirely new. For the ultimate in relaxation, how about a cozy fire while you lounge on the couch and gaze at the stars through the French doors?

surveyor can guide your decision on how to position the house so that water drains away from the house. He can also help determine best positioning for solar energy considerations. He can install your *batter boards* now or at any time prior to number 6. If they may become damaged during steps 3–5, it might be best to wait.

3. **Clearing and excavation.** If your lot is heavily wooded, be sure you clear enough area around the house so that there is enough space for tractors and forklifts to operate. In some areas, local codes require a certain amount of cleared space. A quick phone call can tell you if this is the case. If you are having a basement dug, you may want your surveyor to supervise to be sure of proper depth. Be sure your contract price with your clearing and excavation subcontractor includes hauling away all trash and debris.

4. **Utilities.** Your subcontractors will need water and electricity, so now is the time to have your *saw service* installed, water connection made, or well drilled. In some cases, this is an excellent time to install a septic system. Check with your subcontractor.

NOTE: BE SURE YOU HAVE ORDERED
YOUR BUILDER'S RISK INSURANCE!

5. **Footings.** Types of footings vary as to locale, but they are usually made of concrete poured in trenches or forms. They can also be in the form of pilings or in a combination of footing and concrete slab, called a monolithic slab. Your plans will show clearly what type of footing you will have. Since footings form the basis of your house, you will want to have this step inspected. If your county doesn't have a footing inspector, hire a professional inspector. You might also want to have your surveyor verify that the footings are the exact dimensions of the house and that they are in the right place, although you could do this yourself. Footing inspectors will check for proper depth and to be sure that the footings are on solid *load-bearing* ground. They also check to be sure that they are below the point where the ground freezes, called the frost line. Footings are inspected before the concrete is poured.

6. **Foundation.** As mentioned earlier, your log home can be built on any type of foundation. The foundation can be made of brick, block, poured concrete, concrete slab, pilings, and all-weather wood foundations. You can have your carpenter or surveyor check the foundation for levelness and squareness if you like. Be sure that the crawl spaces are high enough. Most codes require a minimum of 18 inches. Be sure that the basement walls are high enough so that you will have sufficient headroom. Be sure that any foundation is high enough so that water can be diverted around it and that no portion of your wood walls is closer than 8 inches to the finished grade.

After the foundation is finished and, in the case of a concrete slab, before the con-

Can the Trucks Get In?

Log homes are often situated in nice wooded settings with winding driveways. Logs for the home are delivered on long, heavy flatbed trucks. Some unhappy homebuilders have had to cut trees at the last moment or have ended up paying for a wrecker service to tow the delivery truck because they didn't allow for the massive size and weight of these delivery vehicles when planning their driveways. One man spent over $1,000 on wreckers to remove a truck from the mud and mulch of the construction site driveway. Plan ahead.

The Wheatland is one of twenty-three standard designs and floor plans offered by Mountaineer Log Homes. It offers a loft overlooking a stone fireplace in the great room and a master bedroom wing with a cathedral ceiling. Mountaineer homes are offered in three different log profiles: round exterior/flat interior; round exterior/round interior; flat exterior/flat interior.

crete is poured, you or your subcontractor need to call your soil treatment company (exterminator) to have the soil treated against termites. The foundation should then be waterproofed, although this step can be done later. Your lender will require a foundation survey to show that the house is not in violation of any setbacks and is of the dimensions indicated on your plans. You should call your surveyor to request this survey immediately after the foundation is completed, because you won't be able to receive any construction funds until your lender has a copy of the survey and a copy of your builder's risk insurance policy.

7. Plumbing rough-in, for slabs. Prior to pouring concrete, your plumber will install any pipes that will be under the slab. His work needs to be inspected before the concrete is poured. Any electrical conduits need to be installed also and inspected prior to pouring. If you don't have inspectors for these two items in your area, hire a professional.

8. **Slabs.** Concrete slabs need to be inspected before the concrete is poured, but after completing number 7. This inspection is to assure that the slab will comply with codes. Most codes require the following: proper packing down of fill dirt, called tamping; 4–6 inches of sand or gravel for drainage; wire mesh; polyurethane (poly); a border of Styrofoam for insulation; a uniform thickness of the slab throughout; and treatment of the soil. Also, the places where the *load-bearing* walls or posts will rest need to have the slab thickened in accordance to code, usually the same thickness as the footing. A good concrete subcontractor will do all of this, even order the soil treatment. If you don't have local inspectors for this step, hire a professional.

9. **Kit construction and drying-in.** As mentioned in chapter 8, it is best if you find a carpenter who has built a log home before, preferably one of the kind you are building. Construction manuals or guides that instruct your carpenter about kit construction are usually available from the manufacturer. If any questions come up, your log home company can answer them and, in some cases, send an expert out to the job if the need arises. Drying-in is the stage of construction where your house is protected from rain or snow. It doesn't mean that the roof shingles are on, but it can. If not roof shingles, building paper (felt) will protect the interior. Exterior trim of the *fascia, soffit,* and *gable* ends should be included in your contract with your carpenter and can be completed at this time.

10. **Chimney and fireplace(s).** Prior to installation of roofing shingles, fireplaces and chimneys should be built or installed. By finishing this step before roofing, proper *flashing* can be installed around the chimney. It also prevents damage to the shingles to have this step completed first.

11. **Roofing.** Roof shingles can be installed earlier, but it is best to wait. If done before chimneys, the roofer should leave a sufficient area around where any chimney masonry is to be built. The roofer will then have to come back later to finish. If so, only pay for the squares installed, and possibly hold back 10 percent of that.

12. **Plumbing, HVAC, and electrical rough-ins.** Usually the electrician waits until the plumbing and HVAC are roughed in before he starts work. This lessens the chance of having his wires cut accidentally. If your blueprints don't have an electrical plan, and many don't, he will go through the house with you and mark where you want switches, special receptacles (outlets), lights, or any other wiring. The same is true of the HVAC.

(Courtesy of Brentwood Log Homes)

This is the Pathfinder by Brentwood Log Homes. This Georgian-style home exemplifies the country farmhouse found across rural America. These 2,174 square feet of spacious living guarantee a lifetime of homespun living pleasure. Four bedrooms plus one full and one half bath provide a family space to grow. The logs used in construction of this home measure 6 by 12 inches with a dovetailed notch. The ceiling beams are 4-by-8-inch hand-hewn timbers. Logs are available up to 40 feet in length and are West Coast hemlock or eastern white pine. Poplar is also available.

You don't want heat vents where you plan to place a piece of furniture. Plumbing is usually done in strict accordance with your plans, but your plumber may want to go through the house with you before he begins to be sure you understand and agree with the plans and specifications. This is wise, because things may look different in reality. Tubs and molded showers are installed during *rough-in*. With all three subcontractors, you can meet at your convenience (i.e., during lunch hours, after work, or on weekends). All three rough-ins require inspections.

13. **Insulation.** Even if your walls don't need insulation, the rest of your house does. Now is the time for it. It should be inspected in compliance with local codes and/or utility company codes. If you don't have a local inspector, often the utility company does. Inspection of insulation checks for proper installation, proper material and thickness, proper vapor barriers, and packing of spaces and cracks around windows and doors, etc.

14. **Hardwood flooring and carpet underlayment.** This step can be done after number 15, but it is easier to do now. NOTE: If your *subfloor* is ¾-inch tongue-and-groove plywood, you won't need carpet underlayment.

15. **Drywall or paneling.** In areas that will have moisture, you may want to use waterproof Sheetrock.

16. **Interior trim.** Doors, moldings, cabinets, bookcases, etc., are now ready to be installed.

17. **Painting and staining.** Your painter can do the exterior staining after number 9, but he may want to wait and do the whole job at one time. Discuss this with him. If you wait too long for the outside work, you could have uneven fading, excessive *checking,* mold, etc. Of course, if you are merely going to treat the wood and let it age or weather, outside staining won't be your concern. You might want, however, to stain the exterior with a weathering stain that promotes even weathering. This would also provide protection to your wood.

18. **Tile, countertops, etc.** At this time countertops and bath floors must be finished so that the plumbing can be completed. Kitchen floors also are finished at this point. You can wallpaper at this time, unless you want to wait either for final selection of color and design or to see how the cost of the house is coming out.

19. **Trim out plumbing.** Also called "setting the fixtures." At this time your plumber will install sinks, commodes, water heater, faucets, dishwasher and disposal (but not the wiring of these appliances), and any other plumbing fixtures called for in your plans and specifications. He needs to precede your electrician. A final plumbing inspection is necessary.

20. **Trim out HVAC.** At this time the heating and air system is finished and inspected. This also needs to be completed before your electrician finishes wiring the house.

21. **Trim out electrical.** The electrician will install switches, receptacles, and *circuit breakers.* He will also wire the appliances installed by the plumber, as well as the furnace(s) and AC units; install and wire electric ovens and ranges and any other electrical appliances or devices according to your plans and specifications; and wire and hang all light

fixtures, including doorbells. A final electrical inspection is necessary.

22. **Floor finish and/or carpet.** Have floors finished before carpeting. You may want to protect all floors when done with red waxed paper available at most building supply companies.

23. **Cleanup.** You can do some or all of this, but doing windows can be dangerous if you have to get on a ladder.

24. **Drives and walks.** This step can be done earlier, but heavy trucks can cause damage.

25. **Landscaping.** Most lenders require landscaping to be completed before they will close the permanent loan. This can be difficult at certain times of the year, so you may want to start on it as early as possible.

26. **Final inspections.** Besides the inspections already mentioned, you may be required to have a final building inspection. This inspection is to assure the safety of the house (e.g., handrails where required on stairs). It also is to be sure that everything meets local and state codes. Your lenders, both construction and permanent, will make a final inspection. If you don't have a local building inspector, for your own peace of mind, hire a professional.

27. **Loan closing.** This can be arranged at a convenient time for you. Your attorney will prepare all the paperwork and inform you if you need to obtain any *lien waivers* or other documents, such as your homeowner's insurance policy. If you do, be sure he tells you a few days in advance. As defined in the glossary, a lien waiver merely proves that any materials or labor have been paid for.

28. **ENJOYMENT FOR A LIFETIME!**

CHAPTER 11

Useful Tips That Save

Buying Land

• Seldom is the asking price for land the same as the selling price. Remember that when you are buying your land, because it could save you a lot of money. Most sellers have a real estate commission built into their asking price as well as a cushion to negotiate. If you are not using a broker, you should ask for at least the amount a broker would earn as a discount. Then try to get the price down further still. Nothing ventured, nothing gained. It's a game, sometimes maddening perhaps, but it's the only game I know of where there can be two winners.

• Before making a final decision on purchasing your land, have your surveyor determine if your home can be situated so that your sewer or septic system will have a gravity flow and not require a lift station or pump. He most likely will have to call in either city sanitation engineers or health officials, or engineers (for specific system location), but let him handle it. Avoiding a lift station is wise. They are expensive, and if they break down or the power goes out, you'll have a problem.

• Also, be sure there will be adequate drainage of surface water, or runoff, away from your house. Your surveyor can determine this quite easily. Natural grades can usually be changed to accomplish this but not always. Changing grades may require a higher and more costly foundation. Also, fill dirt and/or more grading will increase costs. Water problems can be avoided more easily than they can be remedied. Builders have a healthy respect for water. You will, too, if you consider the Grand Canyon.

• If your house is to be built on a crawl space or slab, try to select land or a lot with a relatively flat building site. This will cut down on foundation costs.

• Wooded land usually has a higher resale value. It is also usually easier to landscape and maintain landscaping by leaving most areas natural.

Planning

• Review your plans very carefully during the planning stage. Try to make any and all changes before you get estimates. Changes later on, especially in the building stage, are very expensive. In my estimation, changes are one of the most significant causes of cost overruns, as well as misunderstandings with suppliers and subcontractors.

- Every reduction you can make in square footage will reduce construction costs.

- Keep roof pitch moderate and roof design simple in order to keep costs down. The higher the pitch, the more valleys and ridges, the higher the cost.

- Laundry rooms can be located right outside bedroom areas (where most dirty laundry accumulates) even in two-story houses. A floor drain should be provided in case of a leak or overflow. In some cases, this move will lower plumbing costs by closer grouping of plumbing (closer to the baths), therefore using less material to plumb. The convenience of having laundry rooms near bedrooms should be obvious.

Financing

- Try to shop for money as you would any product. Half a percentage point in interest or closing costs is a lot of money. You can do it over the phone on your lunch hour. Try to get loans with those lenders that offer the best rates.

Suppliers

- Some of the items that you select to go into your house may not be stocked by local suppliers and may take weeks, or even months, to get. Very typical of this situation are plumbing fixtures, lighting fixtures, and other specialty items. Shop and order early to avoid delays. Keep the job moving, and you'll keep the cost of construction interest down.

- On the same subject, don't spend too much time and energy (or money) worrying about the small things that will go into your house. I've seen people agonize over decisions about doorknobs, faucets, etc., only to forget what they look like a few months after their house was completed.

Building

- If you encounter any difficulties in obtaining permits from building inspection agencies due to their unfamiliarity with log homes, contact your log home manufacturer or representative. It is wise to contact the inspection agency when you first start thinking of building a log home. Energy efficiency is usually the item unfamiliar to these agencies. Your log home representative should provide sufficient information to educate them.

- I have found a brick mason subcontractor who stakes the house, digs and pours the footing, installs the batter boards, and lays the foundation. The cost is about the same as using separate subcontractors for each function, but the time to complete the foundation stage is lessened because only one subcontractor needs to be scheduled. It also saves me time for the same reason. I do have my surveyor check for accuracy.

- After clearing your lot, you may want your clearing and excavation subcontractor to spread unwashed crushed stone on your driveway. This will provide a hard surface that will allow access to your job site in wet weather, thereby preventing delays in deliveries.

- When clearing a wooded lot, either give the wood to your subcontractor in return for a lower contract price or have him leave it in long lengths out of the way of the job site for you to cut at your leisure. Don't have him include cutting it into fireplace lengths unless you are willing to pay dearly for it.

- Be sure to have your plumber protect water lines from freezing. It will save you much aggravation and money in the dead of a hard winter. Try to have water lines plumbed into interior walls and well protected in crawl spaces. Usually, in a log home, pipes are plumbed into interior walls because of the difficulty of plumbing them through the logs. But watch for non-log walls, such as *dormers,* garages, and attic areas. Wells should also be protected. Consult with your plumber on all of the above.

- Insulation in non-log walls, as well as *gables,* crawl spaces, and, most importantly, attics and roofs, should be kept as free of moisture as possible. Otherwise, you will have higher energy bills. Good ventilation in attic areas, roofs, and crawl spaces and a moisture barrier in stud walls will aid in keeping moisture at a minimum in the insulation. Keep attic vents open in the winter. Crawl space vents, because of the prospect of having water lines freeze, unfortunately can't be left open.

- In a house with upstairs plumbing, I recommend cast-iron drain lines between the first and second levels. This cuts down on noise considerably. The rest of the drains can be polyvinyl chloride (PVC) to cut costs. Be sure this is indicated in your specifications.

- Interior bathroom walls can be insulated to cut noise. It's not too expensive.

- You may be able to get a lower price from your painting subcontractor if he can prime coat any interior walls that are to be painted before your carpenters install the interior trim. This is especially true if the interior trim is to be stained. If he can stain the trim before it is installed, a further savings might be realized.

- For log homes requiring chinking, new materials have revolutionized the chinking process. The materials used look like traditional mortar chinking, but while maintaining a strong adhesive bond, they remain flexible. Since a log home is subject to shrinkage and movement, as is any home, in the past chinking would crack and fall out. But with the new materials, the chinking stays in place. Two companies that market these materials are Perma-Chink and Weatherall.

APPENDIX

(Courtesy of Maple Island Log Homes)

This solid oak log home features a loft overlooking a cathedral ceiling in the living room.

Precut Log Home Manufacturers

As published in *Log Home Living* magazine's *Annual Buyer's Guide*

Acadiana Log Homes
296 DeGeyter Road, #100
Breaux Bridge, LA 70517
(318) 332–4099

A Corp. Log Homes
2800 Patton Way
Bakersfield, CA 93308
(805) 589–2628

Aero Log Homes
P.O. Box 27
Wallace, ID 83873
(208) 753–7771

Affordable Log Homes & Builders Supply
61 Water Street
Maryville, NY 14757
(716) 753–7600

Air-Lock Log Co., Inc.
P.O. Box 2506,
South Highway 85
Las Vegas, NM 87701
(505) 425–8888; (800) 786–0525

Alaska Log of Coeur d'Alene
P.O. Box 2135
Hayden Lake, ID 83835
(208) 772–9723

Algonquin Log Homes
Rural Route 1
Norland, ON
Canada K0M 2L0
(705) 454–2311

Allegany Log Homes
Route 19, Dept. HBP

Houghton, NY 14744
(716) 567–2583

Allpine Log Homes
Route 3, Box 434
Berryville, AR 72616
(501) 253–6709

All Weather Log Homes
4220 East 653 North
Rigby, ID 83442
(208) 745–9459

Alpine Homes, Inc.
East 3441 South Apple Tree Lane
Waupaca, WI 54981
(715) 258–9068

Alpine Log Mill
2640 South 1500 East
Vernal, UT 84078
(801) 781–2651

Alta Industries, Ltd.
Route 30, Box 88
Halcottsville, NY 12438
(914) 586–3336

American Heritage Log Homes
133 Soco Road, Highway 19
Maggie Valley, NC 28751
(704) 926–3411

American Log Homes/Log Home Corp.
P.O. Box 7211
Pueblo West, CO 81007
(719) 547–2135

American Southwest Log Homes
P.O. Box 1360
Pagosa Springs, CO 81147
(970) 264–4176, ext. 71;
(800) 999–0832

American Timbercraft Homes
2331 North 1350 West
Ogden, UT 84404
(801) 782–0811

American Wood House Inc.
1899 Rockvale Road
Lancaster, PA 17601
(717) 464–3862

Amerlink Ltd.
P.O. Box 669
Battleboro, NC 27809
(919) 977–2545

Anthony Log Homes
P.O. Box 1877
El Dorado, AR 71730
(501) 862–3414

Anthony Log Homes
433 Deer Haven
Henderson, NC 28791
(704) 692–9966

Appalachian Log Homes
11312 Station West Drive, Dept A
Knoxville, TN 37922
(423) 966–6440
http://www.alhloghomes.com

Appalachian Log Structures
P.O. Box 614
Ripley, WV 25271
(304) 372–6410; (800) 458–9990
http://www.applog.com

Ark Homes, Inc.
5726 Wedgewood Drive
Little Suamico, WI 54141
(414) 826–5901

Arrowhead Log Homes
W13811 County Highway A
Bowler, WI 54416
(715) 793–5730

Asperline Log Homes
Rural Route 1, Box 240
Lock Haven, PA 17745
(717) 748–1880; (800) 428–4663
asperlin@oak.kcsd.k12.pa.us

Authentic Log Homes, Inc.
19 Sandcreek Road
Laramie, WY 82070
(307) 742–3786

Authentic Log Homes, Inc.
Route 14
Hardwick, VT 05843
(802) 472–5096

Back Home to Logs, Inc.
Rural Route 1, Box 112A
Parish, NY 13131
(315) 625–7191

B & B Log Homes/Products
20619 Dunham Road
Marengo, IL 60152
(815) 943–4211

B & H Millwork, Inc.
P.O. Box 5314
Falmouth, VA 22403
(540) 752–2480

Battle Creek Log Homes
9955 Ladds Cove Road
South Pittsburg, TN 37380
(423) 837–7268

Bear Creek Log Homes, Inc.
1660 Little Bear Road
Gallatin Gateway, MT 59730
(406) 763–4709

Bear River Log Homes
567 South Highway 33
Victor, ID 83455
(208) 787–2946

Beaver Log Homes
P.O. Box 236
Beloit, WI 53512
(608) 365–6833

Beaver Log Homes, Inc.
P.O. Box 3
Kalkaska, MI 49646
(616) 258–5020

Beaver Mountain Log Homes
R.D. 1, Box 32
Hancock, NY 13783
(607) 467–2700/2758

Best Log Industries, Inc.
25365 County Road H
Cortez, CO 81321
(970) 565–3932

B.K. Cypress Log Homes, Inc.
609 Gilbert Street
Bronson, FL 32621
(352) 486–2470

Black River Country Log Homes
N7415 West Snow Creek Road

Black River Falls, WI 54615
(715) 284–2893

Blue Moon Co.
P.O. Box 1566
Highlands, NC 28741
(704) 526–3905

Blue Pine Log Homes
P.O. Box 1540
Lander, WY 82520
(307) 332–4542

BPB Log Homes
P.O. Box 86
Goshen, VA 24439
(540) 997–9251

Brentwood Log Homes
201 Walker Springs Road
Knoxville, TN 37923-3106
(615) 982–3788

Bridger Mountain Log Homes
P.O. Box 88
Belgrade, MT 59714
(406) 388–2030

Cal Cedar Homes, Inc.
2160 Green Hill Road
Sebastopol, CA 95472
(707) 829–1511

Canadian Cedar Log Ltd.
3801 19th Street NE
Calgary, AB; Canada T2E 6S8
(403) 291–6465; (800) 346–9291

Canalog Wood Industries Ltd.
P.O. Box 740
Cranbrook, BC
Canada V1C 4J6
(604) 489–5200

Carolina Cabins & Stonework
Route 2, Box 826
Elkin, NC 28621
(910) 874–7680

Carolina Pines Log Homes
P.O. Box 341
Edneyville, NC 28727
(704) 685–7560

Cedar Craft Log Homes
5409 Ely Highway
Middleton, MI 48856
(517) 236–7395

Centennial Log Homes
P.O. Box 5100
Kalispell, MT 59903
(406) 756–6502

Century Cedar Log Homes, Inc.
P.O. Box 746
Running Springs, CA 92382
(909) 867–2421

Cheyenne Log Homes
P.O. Box 1240
Eagar, AZ 85925
(602) 649–5825

Coast to Coast Log Homes
7015 Palmer Road
Millersport, OH 43046
(614) 833–4777; (800) 360–9777

Colorado Log Systems
P.O. Box 8338
Durango, CO 81310
(970) 385–7696

Comins Creek Timber Structures
1166 East Miller Road

Fairview, MI 48621
(517) 848–5948

Coulee Region Log Homes
C.T.V. W5448
Holmen, WI 54636
(608) 526–4421

Country Autumn Log Homes
7687 Highway 135
New Salisbury, IN 47161
(812) 347–3152

Country Log Homes
79 Clayton Road
Ashley Falls, MA 01222
(413) 229–8084

Coventry Log Homes
161 Central Street
Woodsville, NH 03785
(603) 747–3485; (800) 308–7505

Crockett Log & Timber Homes
35 Old Route 12 North
Westmoreland, NH 03467
(603) 399–7725; (800) 566–7714
crockett@top.monad.net

Curr Logging
P.O. Box 13
Chester, ID 83421
(208) 624–7498

Custom Log Homes
P.O. Box 146
Gate City, VA 24251
(540) 386–9495

Custom Log Homes, Inc.
1321 Oak Hill Drive
Clarksville, TN 37040
(615) 552–3574

Custom Nature Homes
2809 Highway 167 North
Lafayette, LA 70507
(318) 232-9568

Deck House
930 Main Street
Acton, MA 01720
(800) 727-3325

Deerfield Log Homes
P.O. Box 1209
Scarborough, ME 04070-1209
(207) 883-5705

Dixie True North
1 Elletson Drive
Greenville, SC 29607
(864) 235-5188

Eagle Ridge Log Homes
P.O. Box 30
Cooks, MI 49817
(906) 644-2600

Edgewood Fine Log Structures
P.O. Box 1030
Coeur d'Alene, ID 83816
(208) 765-9109
http://netnow.micron.net/~edgewood

1867 Confederation Log Homes
Rural Route 3, Box 9
Bobcaygeon, ON
Canada K0M 1A0
(705) 738-5131

El Dorado Log Homes
P.O. Box 212
Greenwood, CA 95635-0212
(916) 888-0725

Enertia Building Systems
13312 Garffe Sherron Road
Wake Forest, NC 27587
(919) 556-2391

Finger Lakes Log Homes
4452 Manning Ridge Road
Painted Post, NY 14870
(607) 936-6636

Fireside Log Homes
P.O. Box 1136
Ellijay, GA 30540
(706) 635-7373

Four Seasons Log Homes
P.O. Box 631
Parry Sound, ON
Canada P2A 2Z1
(705) 342-5211

Frontier Log Homes
188 Hershey Road
Shippensburg, PA 17257
(717) 532-4882

Frontier Log Homes, Inc.
60813 Maple Grove Road
Montrose, CO 81401
(303) 249-7130

Garland Homes/Bitterroot Pre-cut
2172 Highway 93 North
Victor, MT 59875
http://www.garlandhomes.com

Gastineau Log Homes, Inc.
10423 Old Highway 54
New Bloomfield, MO 65063
(573) 896-5122, ext. 246; (800) 654-9253

Genesee Valley Log Homes
3518 Fowlerville Road
Caledonia, NY 14423
(716) 226–3810

Genuine Log Homes
1125 West Baseline Road, Ste. 2-8
Mesa, AZ 85210
(602) 644–0100

George R. White Lumber Co.
P.O. Box 37, Main Street
East Waterford, PA 17021
(717) 734–3816

Georgia Cypress Log Homes
3019 Juanita Lane
Powder Springs, GA 30073
(770) 943–7041

Glacier Log Homes, Inc.
5560 Highway 93 South
Whitefish, MT 59937
(406) 862–3562

Glu-Lam-Log, Inc.
2872 Highway 93 North
Victor, MT 59875
(406) 777–3219

Gold Country Log Homes, Inc.
200 South Cluff Avenue
Lodi, CA 95240
(209) 333–1650

Golden Log Homes by CMB, Inc.
5353 West 56th Avenue
Arvada, CO 80002
(303) 420–2900

Gold Hill Log Homes
P.O. Box 366
Gold Hill, NC 28071
(704) 279–7850

Great Bear Log Homes
P.O. Box 541
Camino, CA 95709
(916) 644–3823

Great Northern Log Homes
7580 Nash Road
Bozeman, MT 59715
(406) 585–9065

Greatwood Log Homes, Inc.
P.O. Box 707, Highway 57
Elkhart Lake, WI 53020
(414) 876–3378; (800) 558–5812

Grenville Log Homes
Rural Route 2
Brockville, ON
Canada K6V 5T2
(613) 925–0508

Harmony Exchange
2700 Big Hill Road
Boone, NC 28607
(704) 264–2314; (800) 968–9663

Heartbilt Homes
1615 Summit Drive
Stockton, IL 61085
(815) 947–3244

Heber Valley Log
2375 South Highway 40
Heber City, UT 84032
(801) 654–5156

Heritage Log Homes, Inc.
P.O. Box 610
Gatlinburg, TN 37738
(423) 436–9331; (800) 456–4663
http://www.heritagelog.com

Hiawatha Log Homes
M-28 East, P.O. Box 8
Munising, MI 49862
(906) 387–4121; (800) 876–8100
http://www.hiawatha.com

Hilltop Log Homes
P.O. Box 270
Bowdoinham, ME 04008
(207) 666–8840

Holland Log Homes Michigan
13352 Van Buren Street
Holland, MI 49424
(616) 399–9627; (800) 968–7564
http://www.hollandloghomes.com

Holmes County Log Homes, Inc.
P.O. Box 220
Berlin, OH 44610
(330) 893–2255

Homestead Log Homes, Inc.
6301 Crater Lake Highway
Medford, OR 97502
(541) 826–6888
http://www.cdsnet.net/Business/Homestead

Homestead Logs, Ltd.
Rural Route 6
Markdale, ON
Canada N0C 1H0
(519) 986–4208

Honest Abe Log Homes, Inc.
3855 Clay County Highway
Moss, TN 38575
(615) 258–3648; (800) 231–3695
http://www.honestabe.com

Honka Southeast
7465 Crosswood Boulevard
Knoxville, TN 37924
(423) 525–1844 ext. 12J

Indus Industries, Inc.
2195 County Road 83 West
Birchdale, MN 56629
(218) 634–2270

Innovations
207 South Kyle Avenue
Republic, MO 65738
(417) 732–5873

International Homes of Cedar, Inc.
P.O. Box 886
Woodinville, WA 98072
(360) 668–8511; (800) 767–7674

Jackson Hole Log Homes
P.O. Box 1747
Jackson, WY 83001
(307) 733–6541

Jim Barna Log Systems, Inc.
2533 North Alberta Street
Oneida, TN 37841
(423) 569–8559

Katahdin Forest Products
Smyrna Road
Oakfield, ME 04763
(207) 757–8278; (800) 845–4533

Kenomee Log Homes, Ltd.
Rural Route 1, Colchester County
Economy, NS
Canada B0M 1J0
(902) 647–2080

Kerry Hix Antique Log Cabins
940 Piney Hill Road
Chatsworth, GA 30705
(706) 695–6431

KSM Enterprises, Inc.
H.C.R. 63, Box 15
McAlisterville, PA 17049
(717) 463–3525

Kuhns Bros. Log Homes
Route 15 South, Rural Route 1, Box 325
Lewisburg, PA 17837
(717) 524–4138
(800) 326–9614
http://www.kuhnsbros.com

Laurentien Log Homes, Ltd.
5636 Route 117
Val-Morin, PQ
Canada J0T 2R0
(514) 229–2933
http://www.loghomes.ca

Lawton Log Homes
P.O. Box 42
Green Valley Lake, CA 92341
(714) 760–8888

Lincoln International
P.O. Box 1177
Woodinville, WA 98072
(206) 885–5383

Lindal Cedar Homes, Inc.
P.O. Box 24426
Seattle, WA 98178
(206) 725–0900, ext. 243-N; (800) 426–0536
http://www.lindal.com

Lodge Logs Homes
3200 Gowen Road
Boise, ID 83705
(800) 533–2450
lodgelog@cyberhighway.net

Logan Log Homes
1501 Smead Road
P.O. Box 878
Logan, OH 43138
(614) 385–6361

Log Cabin Shop and Supply
16547 Meadows Road
White City, OR 97503
(541) 826–6832

Log Creations
S88 W22565, Milwaukee Ave.
Big Bend, WI 53103
(414) 662–2666

Log Forms, Inc.
32 Old Concord Road
Henniker, NH 03242
(603) 428–7776

Log Homes Cooperative of America
P.O. Box 5075
Banner Elk, NC 28604
(704) 758–7777; (800) 564–8496

Log Homes of Louisiana
3108 West Esplanade Avenue
Matairie, LA 70002
(504) 831–2658

Log Homes of the West
P.O. Box 19103
Las Vegas, NV 89119
(702) 735–5647

Log/Panel Systems, Inc.
Route 2, Box 22451
Madisonville, TN 37354
(614) 442–2799

Lok-N-Logs, Inc./Home Office
P.O. Box 677
Sherburne, NY 13460
(607) 674–4447; (800) 343–8928

Lonetree Log Homes
P.O. Box 705
Cedar City, UT 84720-0705
(801) 586–6023; (888) 656–5647

Louisiana Log Home Co., Inc.
1200 I-10 Service Road
Slidell, LA 70461
(504) 649–7091

Lumberjack Log Homes
70 Williams Road
Gallatin Gateway, MT 59730
(406) 763–4421

Maine Cedar Log Homes
35 Main Street
South Windham, ME 04062
(207) 892–8561; (888) 656–5647
http://www.ted.net/~lonetree

Maine Pine Log Homes
c/o Hammond Lumber, Route 27
Belgrade, ME 04917
(207) 495–3303

Mallard Pond Log Homes
Rural Route 2, Box 142–D

Alfred, ME 04002
(207) 324–5108

Mann Made Log Homes
P.O. Box 76
Big Fork, MT 59911
(406) 837–1155

Maple Island Log Homes
2387 Bayne Road
Twin Lake, MI 49457
(616) 821–2151
http://www.mapleisland.com

Mark IV Development, Inc.
P.O. Box 888
Buffalo, WY 82834
(307) 684–2445

Meadow Valley Log Homes
P.O. Box 16, State Highway 173
Mather, WI 54641
(608) 378–4024

Medicine Creek Log Home, Inc.
P.O. Box 308
Medicine Park, OK 73557
(405) 529–2766

Midwestern Log Homes
P.O. Box 188
Hillsdale, MI 49242
(800) 272–9502

Midwest Log Homes Enterprise
705 Chestnut Court
Algonquin, IL 60102
(708) 658–4440

Model Log Homes
75777 Gallatin Road
Gallatin Gateway, MT 59730
(406) 763–4411

Modern Log Homes
1984 Linn Street
Kansas City, MO 64117
(816) 221–5009

Montana Timber Structures
P.O. Box 429
Corvallis, MT 59828
(406) 961–4469

Moose Creek Log Homes
P.O. Box 204L
Turner, ME 04282
(207) 224–7497

Moosehead Cedar Log Homes
P.O. Box 1285
Greenville, ME 04441
(207) 695–3730

Mooselook Ridge Log Homes
P.O. Box 353
Dover-Foxcroft, ME 04426
(207) 564–8665

Mountaineer Log & Siding Co.
P.O. Box 570
McHenry, MD 21541
(301) 387–9200; (800) 336–LOGS
logs@gcc.cc.md.us

Mountaineer Log Homes, Inc.
P.O. Box 248
Morgantown, PA 19543
(610) 286–2005
MTLOGHOMES@aol.com
http://www.cynet.net/logo/index.html

Mountain Gem Log Homes
Box 0038
Laclede, ID 83841
(208) 263–3867

Mountain Log Home Co.
13080 Highways 32 & 64
Mountain, WI 54149
(715) 276–3003

Mountain Log Homes Corp.
Route 1, Box 198
North Bangor, NY 12966
(518) 483–0846

Mountain State Log Homes
Route 2, Box 6AA
Ireland, WV 26376
(304) 452–8228

Mountain Timber Products
11170 West Highway 50
Poncha Springs, CO 81242
(719) 539–4929

Mountain Valley Log Homes
135 South Main, Ste. 101
Heber City, UT 84032
(801) 654–5791

Mountainview Log Structures
P.O. Box 579
Lumby, BC
Canada V0E 2G0
(604) 547–9746

Mount Creek Homes
2190 Cotze Creek Road
Colville, WA 99114
(509) 732–4017

National Log Homes
P.O. Box 2370
Missoula, MT 59806
(406) 542–8809

Nature Log Homes
5685 U.S. Highway 84 West
Mount Enterprise, TX 75681
(903) 863–5665

NettieBay Log Homes
37053 Highlite
Sterling Heights, MI 48310
(810) 979–4666

Neville Log Homes
2036 Highway 93 North
Victor, MT 59875
(406) 642–3091

New Beginnings Log Homes, Inc.
2445 Highway D
Marthasville, MO 63357
(314) 433–5579

New Homestead Log Homes
P.O. Box 61
Creston, IA 50801
(515) 782–2890

New Pioneer Log Homes
3048 Wilson Road
Wieppe, ID 83553
(208) 435–4592

Newton Log Homes
P.O. Box 749
Newton, TX 75966
(409) 379–3937

North American Logs, Inc.
P.O. Box 20494
Bloomington, MN 55420
(612) 884–6902

North American Solid T-F Homes
289 North Main Street

Holland, NY 14080
(716) 537–2320

North Country Log Homes
P.O. Box 1180
Chapleau, ON
Canada P0M 1K0
(705) 864–1190
rkorpela@cancom.net

Northeastern Log Homes, Inc.
P.O. Box 46
Kenduskeag, ME 04450
(207) 884–7000; (800) 624–2797
http://www.northeasternlog.com

Northern Land & Lumber Co.
7000 P Road
Gladstone, MI 49837
(906) 786–2994

Northern Log Homes
2381 Old Keene Road
Athol, MA 01331
(508) 249–9439

Northern Log Homes
#6 2789 Highway 97 North
Kelowna, BC
Canada V1X 4J8
(604) 765–2408

Northern Log Homes
300 Bomarc Road
Bangor, ME 04401
(207) 942–6869

North Star Log Homes & Lumber Co.
W3516 LaBelle Road
Powers, MI 49874
(906) 497–5020
http://www.tm/arts.com/nstar/nstar.htm

Northwoods Log Homes, Inc.
P.O. Box 645
LaPorte, MN 56461
(218) 224–2251

Old Mill Log Homes
HC 89, Box 115B
Pocono Summit, PA 18346
(717) 839–1445

Old Timer Log Homes & Supply
1901 Logue Road
Mount Juliet, TN 37122
(615) 443–0080; (800) 321–5647
http://www.oldtimerloghomes.com/~bestlogs

Original Lincoln Logs Ltd.
P.O. Box 135 Riverside Drive
Chestertown, NY 12817
(518) 494–5500; (800) 833–5500
http://www.lincolnlogs.cominfo
@lincolnlogs.com

Original Log Cabin Homes
P.O. Drawer 1457
Rocky Mount, NC 27802
(919) 977–7785; (800) 56CABIN
http://www.logcabinhomes.cominfo
@logcabinhomes.com

Original Log Cabins Ltd.
Box 239
Pine River, MB
Canada R0L 1M0
(204) 263–5209

Ostego Cedar Log Homes
P.O. Box 127
Waters, MI 49797
(517) 732–6268

Outaouais Log Homes
Box 157

Wakefield, PQ
Canada J0X 3G0
(819) 459–2089

Pacific Log Homes Ltd.
P.O. Box 80868
Vancouver, BC
Canada V5H 3Y1
(800) 663–1577
pacific@lightspeed.bc.ca

Pan Abode Cedar Homes, Inc.
4350 Lake Washington Boulevard North
Renton, WA 98056
(206) 255–8260

Panabode International Ltd.
6311 Graybar Road
Richmond, BC
Canada V6W 1H3
(604) 270–7891

Pariso Log Homes
4856 State Route 39 NW
Dover, OH 44622
(330) 343–7320

Patriot Log Home, Ltd.
15 Robert Street
St. Sauveuer des Monts, PQ
Canada J0R 1R6
(514) 227–4608

Peace Country Log Builders
Box 992
Beaverlodge, AB
Canada T0H 0C0
(403) 356–2614

Pinecraft Log Homes, Inc.
2805 County Road JJ
Neenah, WI 54956
(414) 729–6132; (800) 729–6132

Pinnacle Custom Log Homes
2646 School Lane
Green Bay, WI 54313
(414) 434–8892

Pioneer Log Systems, Inc.
181 West Kingston Springs Rd.
Kingston Springs, TN 37082
(615) 952–5647

Precision Craft Log Structures
711 East Broadway
Meridian, ID 83642
(208) 887–1020; (800) 729–1320
http://www.precisioncraft.com

Proctor Piper Log Homes
R.F.D. 1, Box 146
Proctorsville, VT 05153
(802) 226–7224

Rapid River Rustic Cedar Homes
P.O. Box 10
Rapid River, MI 49878
(906) 474–6404; (800) 422–3327
http://visitusa.com/loghome

Raystown Land Company
R.D. 1, Box 29
James Creek, PA 16657
(814) 658–3469

Real Log Homes/Vermont Log Buildings
P.O. Box 202
Hartland, VT 05048
(802) 436–2130
(800) REAL LOG (732–5564)

Riverbend Log Homes
P.O. Box 411
Nackawic, NB
Canada E0H 1P0
(506) 575–2719

Rocky Mountain Log Homes
1833 L Highway 93 South
Hamilton, MT 59840
(406) 363–5680

Roger Simota Country Homes
R.R. 1
Trempealeau, WI 54661
(608) 534–6414

Rogers Wood Products
R.D. 1, Box 117
Mooers, NY 12958
(518) 236–7574

Ross-American Hardwood, Inc.
7 East Lincoln Avenue
Lake Wales, FL 33853
(941) 678–3325

Rustic Elegance Log Homes
6303 Rich Road
Olympia, WA 98501
(360) 493–1284

Salisbury Hardware & Lumber
5700 U.S. Highway 31
Grawm, MI 49637-9652
(616) 929–4211

Satterwhite Log Homes
Route 2, Box 256A
Longview, TX 75605
(903) 663–1729
(800) 777–7288
http://www.dfw.net/~loghome/

Schweizer Lumber Co.
R.D. 1, Route 162
Sloanesville, NY 12160
(518) 895–8081

Shenandoah Log Homes
P.O. Box 590
Canadensis, PA 18325
(717) 223–9270

Sierra Log Homes
2484 Honey Run Road
Chico, CA 95928
(916) 899–0680

Sing Square Log Homes
P.O. Box 11532
Bainbridge Island, WA 98110
(206) 780–1751

Snake River Log Homes
4220 East 653 North
Rigby, ID 83442
(208) 745–9459
loghomes@srv.net

South Eastern Log Homes Manufacturing
408 Putnam Road
Fountain Inn, SC 29644
(800) 847–5647

Southern Cypress Log Homes
U.S. Highway 19 South
Crystal River, FL 34423
(352) 795–0777

Southland Log Homes, Inc.
Route 2, Box 1668
Irmo, SC 29063
(803) 781–5100; (800) 845–3555

Stonemill Log Homes
7015 Stonemill Road
Knoxville, TN 37919
(423) 693–4833; (800) 438–8274
stonemill@ix.netcom.com

Straight River Log Homes
Route 3, Box 1A
Park Rapids, MN 56470
(218) 732–5638

Sula Log Homes
Highway 93 South, Box 5679
Conner, MT 59827
(406) 821–3831

Summit Log Homes
P.O. Box 185
Blowing Rock, NC 28605
(704) 266–1102

Superior Log Homes, Inc.
5253 Corey Road
Williamston, MI 48895
(517) 468–3344

Suwannee River Log Homes
P.O. Box 610
Wellborn, FL 32094
(904) 963–5647
(800) 962–LOGS (5647)

Tennessee Log Homes, Inc.
2537 Decatur Pike
Athens, TN 37303
(423) 745–8993; (800) 251–9218
http://www.tnloghomes.com/tnlog

Teton Peaks Log Homes, Inc.
4080 East 600 North
Rigby, ID 83442
(208) 745–8089

Thomas Service Co., Inc.
Route 1, Box 548
Mooreville, MS 38857
(601) 842–0966

Timber Lifestyles of Montana
511 Franklin
Hamilton, MT 59840
(406) 363–1070

Timberline Log Homes
6002 East Greenway Lane
Scottsdale, AZ 85254
(602) 951–9404

Timber Log Building Systems
639 Old Hartford Road
Colchester, CT 06415
(860) 537–2393; (800) 533–5906

Timber Ridge Log Homes
2302 Walters Road
Allison Park, PA 15101
(412) 487–4047

Timberstone Log Homes
Rural Route 1
Berwick, NS
Canada B0P 1E0
(902) 538–7898

Timber Structures
P.O. Box 19446
Greensboro, NC 27419
(910) 664–1247

The Timber Touch
P.O. Box 1542
Nevada City, CA 95959
(916) 265–4997

Timberyard Log Homes
27 Archie Lane
Pine Mountain, GA 31822
(706) 663–8477

Tomahawk Log & Country Homes
2285 Business 51 County L

Tomahawk, WI 54487
(715) 453–3265; (800) 544–0636

Town & Country Cedar Homes
4772 U.S. 131 South
Petoskey, MI 49770
(616) 347–4360; (800) 968–3178
http://www.cedarhomes.com

Traverse Bay Log Homes
6446 East Traverse Highway (M-72W)
Traverse City, MI 49684
(616) 947–1881

Tree Craft Log Homes, Inc.
P.O. Box 34
Mars Hill, NC 28754
(704) 689–2240

Treehouse Log Homes
Route 5, Box 315
Park Rapids, MN 56470
(218) 732–8092

T.R. Miller Mill Co., Inc.
P.O. Box 708
Brewton, AL 36427
(205) 867–4331

True-Craft Log Structures, Ltd.
25352 Lougheed Highway
Box 129
Whonnock, BC
Canada V0M 1S0
(604) 462–8833

True North Log Homes
P.O. Box 2169
Bracebridge, ON
Canada P1L 1W1
(705) 645–3096; (800) 661–1628
trunorth@mustcoka.com

Tussey Mountain Log Homes, Inc.
R.D. 1, Box 53A
Pittsfield, PA 16340
(814) 563–4648

Unique Timber Corp.
P.O. Box 730
1837 Shuswap Avenue
Lumby, BC
Canada V0E 2G0
(250) 547–2400
http://www.uniquetimber.com

Voyageur Log Homes, Inc.
5228 Old Highway 53
Orr, MN 55771
(218) 757–3108

Wahconah Log Homes
75 Merrills Ridge Drive
Asheville, NC 28803
(704) 299–1136

Ward Log Homes
P.O. Box 72
Houlton, ME 04730
(207) 532–6531
(800) 341–1566

Western Homes Corp.
2322 Boulder Bluff
Riverside, CA 92506
(909) 734–6610

Weyerman's Log Mill, Inc.
13250 North Dry Fork Canyon
Vernal, UT 84078
(801) 789–5647

Whispering Pines Log Homes
Box 99, Highway 10 West
Verndale, MN 56481
(218) 631–1974

Wholesale Log Homes, Inc.
P.O. Box 177
Hillsborough, NC 27278
(919) 732–9286

Wholesale Logs & Lumber
R.D. 2, Box 79
Scottdale, PA 15683
(412) 887–3020

Wilderness Building Systems
3015 South 460 West
Salt Lake City, UT 84115
(801) 972–6066

Wilderness Log Homes, Inc.
P.O. Box 902-LH
Plymouth, WI 53073
(800) 237–8564
http://www.cae.wisc.edu/~sunf/log/loghome.html

Wisconsin Log Homes, Inc.
P.O. Box 11005
Green Bay, WI 54307
(414) 434–3010; (800) 844–7970

Wooden Indian Log Homes Co.
2101 Old Lancaster Pike
Reinholds, PA 17569
(610) 796–9663

Woodland Log Home Supply
6850 West State Road 46
Mims, FL 32754
(407) 269–1213

Yellowstone Log Homes
280 North Yellowstone
Rigby, ID 83442
(208) 745–8108

Internet Sites

Allegany Log Homes
www.rctc.com/aloghome/kilndried.html

Anthony Log Homes
www.anthonyforest.com/loghome.html

Appalachian Log Homes
www.alhloghomes.com/main.phtml

Barna Log Homes of Michigan
www.barnalogs.com

Beaver Mountain Log Homes
www.beavermtn.com/

Black River Country Log Homes, Inc.
www.r2j2.com/barry/gif05.html

Brown Bear Log Homes
www.value-link.com/brownbear.html

Caribou Creek Log Homes
www.caribou-creek.com

Cedar Homes
www.cedarhomes.com

Confederation Log Homes
www.jurock.com/buildworks/
confederation.html

DST Construction—Amerlink
www.dstlog.com

Fireside Homes
www.firesidehomes.com

Garland Log Homes
www.garlandhomes.com/

Hiawatha Log Homes
www.hiawatha.com

Holland Log Homes
http://hollandloghomes.com

Homestead Log Homes
www.cdsnet.net/Business/Homestead/index.
htm

Honest Abe Log Homes
http://honestabe.com

Kuhns Brothers Log Homes
www.Kuhnsbros.com/

Lincoln Log Homes
www.lincolnlogs.com

Linwood and Pan-Abode Cedar Homes
www.pbi-hawaii.com/faq.htmm

Northeastern Log Homes
www.northeasternlog.com/

Northeastern Log Homes
www.mainemade.com/h/h 06.html

Norse Log Homes
www.norse-log-homes.com

Northern Log Homes
www.mainerec.com/loghomes/index1.html

Precision Craft Log Homes
www.precisioncraft.com

Pacific Log Homes
http://lightspeed.bc.ca/warlight/PLH/
pacific.html

Riverbend Log Homes
www.lexicom.ab.ca/~riverbend/cabin.htm

Rocky Mountain Log Homes
www.Rocky-mtn-log-homes.com

Satterwhite Log Homes
www.satterwhite-log-homes.com/

Sentry Handcrafted Log Homes
www.boreal.org/sentry/index.html

Soda Creek Log Homes
www.cariboo-net.com/sodack.html#
anchor2157895

Southland Log Homes
www.intraweb.com/loghomes.htm

Stone Mill Log Homes
www.smokymtnmall.com/mall/stoneml.html

Summit Log Homes
www.appnetsite.com/summit.htm

Tennessee Log Homes
www.tnloghomes.com

Timberline Log Homes
www.montana.com/ibs/timber/timber.html

Glossary

Backfilling—Filling an area with dirt, sand, or stone to bring it up to desired grade (level).

Batter boards—Boards erected to show the proper height and corners of a foundation.

Bids—The amount of money for which a subcontractor is willing to do a specific job. Also called quotes.

Building codes—Sets of laws that establish minimum standards in construction. Codes and enforcement of compliance vary with locale.

Capping—Covering the ridge of a roof with roofing material.

Certificate of insurance—Proof of insurance.

Checking—Surface cracks in wood.

Chinking—Closing up the gaps between the logs in a chink-style log home.

Circuit breakers—Electrical devices that prevent overloading of an electrical circuit.

Clear title—Proof of ownership of any property that is free of any encumbrances, such as liens, mortgages, and judgments.

Collateral—Physical property pledged as repayment of a loan.

Contracts—Quotes or bids agreed upon by you and the subcontractor and put in writing.

Deed—A legal document that transfers ownership of property.

Dormer—A window, or even a room, that projects from a sloping roof.

Draws—Disbursements of money that equal a percentage of the total amount due.

Drying-in—Term indicating that your house is far enough along as to be protected from rain or snow. Also called "in the dry" and "weather tight."

Expansion joint—A joint in concrete to allow for expansion of the concrete with temperature changes. It is usually made of fiberboard or Styrofoam.

Fascia—Also called fascia board. The trim board around a roof's edge.

Flashing—Sheets of metal used to weatherproof roof joints.

Footings—The base of a structure, usually made of concrete, that supports the foundation of a house.

Framing members—The materials used to put the house together, other than the logs.

Gable—The triangular wall formed by the sloping ends of a ridged roof.

Heat gain—A term used to describe heat entering a building.

Heat loss—A term used to describe heat escaping from a building.

Letter of commitment—A letter from a permanent lender stating that it will make a permanent loan to you, provided that all the conditions that led it to this decision are the same at the time of closing. Its purpose is to allow you to obtain construction financing.

Lien waiver—A legal document that states that an individual or firm has been paid in full for the labor or supplies that went into your home.

Load-bearing—Capable of carrying the weight of a structure.

Lot subordination—A process of buying land whereby the entire purchase price does not have to be paid in order to receive a deed to the land. The seller takes a promissory note or a (second) mortgage and subordinates his rights to those of the construction loan lender.

Manager's contract—A contract with a professional general contractor whereby he will manage as much of and as many of the phases of construction as you wish, with you remaining as the overall general contractor.

Plot plan—A survey of your land showing where the house will be positioned.

Points—A charge for lending money.

Qualify—A term that means one can afford a mortgage.

Quotes—See *Bids.*

Recorded—On file at the local county courthouse, usually attached to a deed.

Recording fees—The fees charged to record a legal document.

Restrictions—Certain restraints placed on a particular lot or parcel of land by the current owner or a previous owner. Size of house, architectural design, number of stories, type of driveway, use of outbuildings, etc., are some of the things covered by restrictions.

Ridge vent—A continuous vent that runs along the ridge (peak) of a roof.

Rough-in—The installation of wiring, plumbing, heat ducts, etc., in the walls, floors, or ceilings (before those walls, floors, or ceilings are covered up permanently).

Sash locks—Window locks.

Saw box—Temporary electrical service receptacle.

Saw service—Temporary electrical service for the purpose of construction.

Sill plate—The horizontal framing member next to the foundation that supports a wall or floor.

Soffit—The underside of a roof's overhang, or cornice.

Subfloor—The floor beneath the finished or final floor. Usually ½-inch plywood.

Survey—A written description of a lot or parcel of land that determines its location and boundaries.

Take-off—An estimate of materials.

Thermal mass—The bulk of logs, stone, or masonry that inhibits the passage of heat through the walls.

Title—Proof of ownership.

Window grids—The mullions or dividers that snap into place to look like windowpanes. Used with insulated glass windows that don't have panes.

Wood foundation—A foundation made of treated wood in lieu of masonry or concrete. Can be installed even in frigid weather.

Zoning—An area of land that is restricted by a local government to a certain use, such as residential, business, or industrial.

INDEX

Note: Boldfaced pages indicate illustrations or charts.

Recommended Reading

Be Your Own House Contractor, Carl Heldmann, Garden Way Publishing, Pownal, VT, 1981.

Build Your Own Low-Cost Log Home, Roger Hard, Garden Way Publishing, Pownal, VT, 1977.

Carpentry and Exterior Finish: Some Tricks of the Trade, Bob Syvanen, The Globe Pequot Press, Old Saybrook, CT, 1993.

Carpentry and Interior Finish: More Tricks of the Trade, Bob Syvanen, The Globe Pequot Press, Old Saybrook, CT, 1993.

Manage Your Own Home Renovation, Carl Heldmann, Garden Way Publishing, Pownal, VT, 1987.

Log Home Magazines & Journals

Country's Best Log Homes
Homestead Communications Corp.
441 Carlisle Drive
Herndon, VA 22070
(703) 471–2041

Log Homes Design Ideas
H&S Media
3400 Dundee Road
Northbrook, IL 60062
(708) 291–1135

Log Homes Illustrated
GCR Publishing Group, Inc.
1700 Broadway
New York, NY 10019
(800) 659–1395

Log Home Living
Home Buyer Publications
4200 T-Lafayette Center Drive
Chantilly, VA 22021
(703) 222–9411; (800) 826–3893
FAX: (703) 222–3209